INTEGRAL
EQUATIONS

INTEGRAL EQUATIONS

F.G. TRICOMI

Dover Publications, Inc.
New York

Published in Canada by General Publishing Company, Ltd., 30 Lesmill Road, Don Mills, Toronto, Ontario.
Published in the United Kingdom by Constable and Company, Ltd., 10 Orange Street, London WC2H 7EG.

This Dover edition, first published in 1985, is an unabridged and unaltered republication of the work first published by Interscience Publishers, Inc., London & New York, in 1957.

Manufactured in the United States of America
Dover Publications, Inc., 31 East 2nd Street, Mineola, N.Y. 11501

Library of Congress Cataloging in Publication Data

Tricomi, F. G. (Francesco Giacomo), 1897–
    Integral equations.

    Reprint. Originally published: New York: Interscience Publishers, 1957.
(Pure and applied mathematics (Interscience Publishers) ; v. 5)
    Bibliography: p.
    Includes index.
    1. Integral equations.   I. Title.
QA431.T73   1985   515.4'5   84-25917
ISBN 0-486-64828-1

Preface

One of the first subjects in mathematics to attract my attention was integral equations; yet this book appears after a score of others. Why is this? It is because the writing of a book on integral equations is a rather difficult task, a task for which many years of meditation are necessary.

In fact, such a book must satisfy two requirements which are not easily reconciled. In order to facilitate theoretical applications to existence proofs it must present the main results of the theory with adequate generality and in accordance with modern standards of mathematical rigor. On the other hand, it must not be written so abstractly as to repel the physicist, engineer, and technician who certainly need and deserve this mathematical tool.

Have I succeeded in satisfying both requirements? Only the reader can decide. I can only hope that, if I have not always been successful in simplifying difficult matters, at least I shall not be found guilty of artificially complicating simple matters, a phenomenon which sometimes occurs in mathematical writing.

In the attempt to reconcile generality with simplicity, I was greatly helped by an idea put forward by my friend, Professor M. Picone.* Although already used by E. Schmidt, one of the founders of the theory of integral equations, this idea is still little known. By means of Neumann's series, it allows one to pass easily from an integral equation with a 'degenerate' kernel to one with a general kernel.

Here and there, especially in the last chapter, the specialist will find a few new things, or some old ones in a new form, e.g. the theory of Volterra integral equations in the L_2-space instead of the space of continuous functions. But, in general, I have avoided modifying material which has already reached a traditional, satisfactory systematization. I believe, moreover, that a book of this character should consider almost exclusively matters and methods which are already well established within the framework of analysis.

* Picone, M., *Appunti di analisi superiore*, Rondinella, Napoli, 1940.

For this reason, I do not use modern topological methods of functional analysis; this may perhaps be done in a future edition. However, some disagreeable exclusions were necessary to avoid too big a volume; for instance, applications of integral equations with any degree of completeness would mean dealing with a large part of mathematical physics and the modern mechanics of vibrations!

This book is meant to be a modern textbook on integral equations for graduate students, applied mathematicians, and so on. For this reason, I have attempted to keep the mathematical knowledge required on the part of the reader to a minimum; a solid foundation in differential and integral calculus and some theory of functions is sufficient.

The book consists of four chapters, each divided into many sections (progressively numbered 1·1, 1·2, ..., 4·7) and two Appendices (I and II).

The formulae are numbered progressively within each section or Appendix and are quoted in the same section with their numbers only. In other sections, this number is preceded by the number of the section. For instance, (4·3–4) denotes (outside of §4·3), formula (4) of §4·3.

I am deeply indebted to my friend, Dr Charles de Prima, who, with his unusual competence in this field, gave many useful suggestions, not only of a mathematical, but also of a linguistic nature. Thanks also go to Miss Ruth Struik who spent a lot of time in polishing the manuscript.

Turin, Italy   F. G. TRICOMI
February, 1955

Contents

CHAPTER I

Volterra Equations

1·1. A Mechanical Problem Leading to an Integral Equation

Integral equations are one of the most useful mathematical tools in both pure and applied analysis. This is particularly true of problems in mechanical vibrations and the related fields of engineering and mathematical physics, where they are not only useful but often indispensable even for numerical computations.

It seems appropriate to illustrate by means of a simple example the intimate connection between the mathematical theory, which forms the subject-matter of this book, and the 'practical' problems of applied sciences. It is well known that, if the speed of a rotating shaft is gradually increased, the shaft, at a certain definite speed (which may at times be far below the maximal speed allowed), will undergo rather large unstable oscillations. Of course, this phenomenon occurs when the speed of the shaft is such that, for a suitable deformation of the shaft, the corresponding centrifugal force just balances the elastic restoring forces of the shaft.

In order to determine the possible 'critical' speeds of the shaft, we may utilize a simple, yet general, result from the theory of elastic beams: For an arbitrary elastic beam under arbitrary end conditions, there always exists a uniquely defined *influence function* G which yields the deflection of the beam in a given direction γ (actually the direction Oz, Fig. 1) at an arbitrary point P of the beam caused by a unit loading in the direction γ at some other point Q. For, if the cross-sections of the beam are placed in one-to-one correspondence with the points of the segment $0 \leqslant x \leqslant 1$, then G is a symmetric† function $G(x, y)$ of the abscissae x and y of P and Q respectively. Consequently by the superposition principle of

---

† That $G(x, y) = G(y, x)$ is a consequence of the Betti-Maxwell reciprocity principle of elasticity.

elasticity, if $p(x)$ is an arbitrary continuous load distribution along the beams† then the corresponding deflection is

$$z(x) = \int_0^1 G(x, y)\, p(y)\, dy \quad (0 \leqslant x \leqslant 1). \tag{1}$$

If we consider the shaft which rotates about the x-axis and in which $z(x)$ represents the deflection of the center of gravity of the cross-

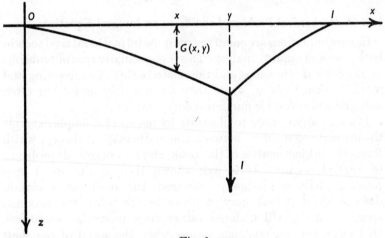

Fig. 1

section corresponding to x, then the load distribution for such a deflection $z(x)$ and an angular speed ω is given by

$$p(x) = \omega^2 \mu(x)\, z(x), \tag{2}$$

where $\mu(x)$ is the linear mass density of the shaft. Hence the balancing of the elastic force and the centrifugal force will occur for a certain ω if and only if there exists a non-zero deflection $z(x)$ which satisfies the equation

$$z(x) = \omega^2 \int_0^1 G(x, y)\, \mu(y)\, z(y)\, dy \quad (1 \leqslant x \leqslant 1). \tag{3}$$

Thus the problem of determining the critical speeds is reduced to determining the values of ω^2 for which the previous equation admits non-zero solutions.

† This signifies that the loading between x and $x + dx$ is $p(x)\, dx$.

This equation is an integral equation; more precisely, it is called a *linear homogeneous Fredholm integral equation of the second kind* with the *kernel* $G(x, y) \mu(y)$.

Using the fact that $\mu(x) > 0$ we can transform advantageously equation (3) into a similar one with a *symmetric* kernel. We set

$$\phi(x) = \sqrt{\mu(x)}\, z(x) \quad \text{and} \quad \omega^2 = \lambda,$$

and immediately obtain equation

$$\phi(x) - \lambda \int_0^1 K(x, y)\, \phi(y)\, dy = 0 \quad (0 \leqslant x \leqslant 1), \tag{4}$$

whose kernel

$$K(x, y) = \sqrt{[\mu(x)\,\mu(y)]}\, G(x, y)$$

is obviously symmetric.

The advantage of this transformation is—as we shall show later (Chapter III)—that a *symmetric* kernel generally possesses an infinity of *eigenvalues* (also called *characteristic* or *proper values*), i.e. values of λ for which the equation has non-zero solutions. On the other hand, a non-symmetric kernel may or may not have eigenvalues.

1·2. Integral Equations and Algebraic Systems of Linear Equations

Let $K(x, y)$ be a given real function defined for $0 \leqslant x \leqslant 1, 0 \leqslant y \leqslant 1$, †
$f(x)$ a given real function defined for $0 \leqslant x \leqslant 1$ and λ an arbitrary complex number; then the general linear *Fredholm integral equation of the second kind* for a function $\phi(x)$ is an equation of the type

$$\phi(x) - \lambda \int_0^1 K(x, y)\, \phi(y)\, dy = f(x) \quad (0 \leqslant x \leqslant 1) \tag{1}$$

while the linear *Fredholm integral equation of the first kind* is given by

$$\int_0^1 K(x, y)\, \phi(y)\, dy = f(x) \quad (0 \leqslant x \leqslant 1). \tag{2}$$

† For the sake of simplicity we shall generally assume the basic domain of our equations to be $0 \leqslant x \leqslant 1$. Only slight changes, if any, will occur if the basic domain is $a \leqslant x \leqslant b$ or even any bounded (measurable) set on the x-axis. In order not to become unduly precise *in this introductory section*, it is implicitly assumed that all functions occurring satisfy conditions which permit us to carry out our operations.

The problem of solving (1) or (2) can be considered as a generalization of the problem of solving a set of n linear algebraic equations in n unknowns:

$$\sum_{s=1}^{n} a_{rs}x_s = b_r, \quad (r = 1, 2, \ldots, n). \tag{3}$$

In fact, if $K(x, y)$ and $f(x)$ are step-functions, i.e. if

$$K(x, y) = k_{rs} \quad \left(\frac{r-1}{n} < x \leqslant \frac{r}{n}, \frac{s-1}{n} < y \leqslant \frac{s}{n}\right)$$

$$f(x) = f_r \quad \left(\frac{r-1}{n} < x \leqslant \frac{r}{n}\right) \qquad (r, s = 1, 2, \ldots, n),$$

$$\tag{4}$$

then equations (1) and (2) become

$$\phi(x) = f_r + \lambda \sum_{s=1}^{n} k_{rs} \int_{(s-1)/n}^{s/n} \phi(y)\, dy \qquad \left(\frac{r-1}{n} < x \leqslant \frac{r}{n}\right), \tag{1'}$$

and

$$f_r = \sum_{s=1}^{n} k_{rs} \int_{(s-1)/n}^{s/n} \phi(y)\, dy \tag{2'}$$

respectively. Equation (1′) shows that if a solution $\phi(x)$ exists it must also be a step-function, i.e.

$$\phi(x) = \phi_r \quad \left(\frac{r-1}{n} < x \leqslant \frac{r}{n}\right);$$

then (1) can be witten in the form

$$\phi_r - \frac{\lambda}{n} \sum_{s=1}^{n} k_{rs}\phi_s = f_r \quad (r = 1, 2, \ldots, n). \tag{1''}$$

Thus, if the determinant with the elements

$$\delta_{rs} - \frac{\lambda}{n} k_{rs} \quad \left(\delta_{rs} = \begin{cases} 0, r \neq s \\ 1, r = s \end{cases}\right)$$

does not vanish, then (1″) and therefore (1′) has a unique solution for any given step-function $f(x)$.

The case of (2′) is different; here one cannot conclude that $\phi(x)$ is necessarily a step-function. All that can be said is that if we set

$$n \int_{(s-1)/n}^{s/n} \phi(y)\, dy = x_s$$

then (2′) becomes

$$f_r = \frac{\lambda}{n} \sum_{j=1}^{n} k_{rs} x_s \quad (r = 1, 2, \ldots, n). \tag{2''}$$

Hence the system $(2')$ in contrast to $(1')$ possesses an intrinsic difficulty: even if the determinant of k_{rs} is non-zero and $(2'')$ has a unique solution, the system $(2')$ will possess *infinitely* many solutions $\phi(x)$, since only the x_r, which are the mean values of $\phi(x)$ in the successive intervals $(0, 1/n)$, $(1/n, 2/n)$, ..., are determined uniquely. This simple analysis merely foreshadows the serious difficulties which may be encountered in a study of integral equations of the *first* kind.

From a slightly different point of view, an integral equation can be considered as a *limiting case* of a system of type $(1')$ or $(2')$.

1·3. Volterra Equations

To avoid some of the difficulties indicated in the previous section, Vito Volterra† investigated the solution of integral equations in which the kernel satisfies the condition

$$K(x, y) \equiv 0 \quad \text{if} \quad y > x. \tag{1}$$

This corresponds (in the sense of the previous section) to the simple case of a system of algebraic linear equations where the elements of the determinant above the main diagonal are all zero. The integral equations

$$\phi(x) - \lambda \int_0^x K(x, y)\, \phi(y)\, dy = f(x) \tag{2}$$

and

$$\int_0^x K(x, y)\, \phi(y)\, dy = f(x) \tag{3}$$

are called *Volterra integral equations* of the second and first kind, respectively. We shall begin our study with this relatively simple,

† The principal founders of the theory of *integral equations* are Vito Volterra (1860–1940) and Ivar Fredholm (1866–1927), together with David Hilbert (1862–1943) and Erhard Schmidt (b. 1876). Volterra was the first to recognize the importance of the theory and to consider it systematically, but Fredholm's contribution consisted in overcoming (about 1900), rather than avoiding, the difficulty connected with the vanishing of the 'determinant of the coefficients.' Nevertheless, Volterra's priority would have been more generally acknowledged if his first paper on the subject (1896) had been presented differently. But Volterra, instead of deducing his results by the same methods he used for their discovery (which were identical with those employed later so successfully by Fredholm) simply published a verification of his solution. This was told to me by Volterra himself when, in 1923–4, I lectured for the first time on integral equations at the University of Rome.

but important, class of equations in which many features of the general theory already appear.

The Volterra integral equation of the second kind† (2) is readily solved by Picard's process of successive approximations. We start by setting $\phi_0(x) = f(x)$ and determine $\phi_1(x)$:

$$\phi_1(x) = f(x) + \lambda \int_0^x K(x, y) f(y) \, dy.$$

Continuing in this manner we obtain an infinite sequence of functions

$$\phi_0(x), \quad \phi_1(x), \quad \phi_2(x), \quad \dots, \quad \phi_n(x), \quad \dots, \tag{4}$$

satisfying the recurrence relations

$$\phi_n(x) = f(x) + \lambda \int_0^x K(x, y) \, \phi_{n-1}(y) \, dy \quad (n = 1, 2, 3, \dots). \tag{5}$$

Better still, setting

$$\phi_n(x) - \phi_{n-1}(x) = \lambda^n \psi_n(x) \quad (n = 1, 2, 3, \dots), \tag{6}$$

we observe that

$$\phi_n(x) = \sum_{\nu=0}^n \lambda^\nu \psi_\nu(x) \tag{7}$$

if $\psi_0(x) = f(x)$, and further that

$$\psi_n(x) = \int_0^x K(x, y) \, \psi_{n-1}(y) \, dy \quad (n = 1, 2, 3, \dots),$$

hence

$$\psi_1(x) = \int_0^x K(x, y) f(y) \, dy$$

and

$$\psi_2(x) = \int_0^x K(x, z) \, dz \int_0^z K(z, y) f(y) \, dy.$$

This repeated integral may be considered as a double integral over the triangular region indicated in Fig. 2; thus, interchanging the order of integration, we obtain

$$\psi_2(x) = \int_0^x f(y) \, dy \int_y^x K(x, z) \, K(z, y) \, dz$$

or

$$\psi_2(x) = \int_0^x K_2(x, y) f(y) \, dy,$$

where

$$K_2(x, y) = \int_y^x K(x, z) \, K(z, y) \, dz.$$

† Volterra integral equations of the first kind will be considered in § 1·6.

Similarly, we find in general

$$\psi_n(x) = \int_0^x K_n(x,y) f(y)\, dy \quad (n = 1, 2, 3, \ldots), \tag{8}$$

where the *iterated kernels*

$$K_1(x,y) \equiv K(x,y), \quad K_2(x,y), \quad K_3(x,y), \quad \ldots$$

are defined by the recurrence formula†

$$K_{n+1}(x,y) = \int_0^x K(x,z) K_n(z,y)\, dz \quad (n = 1, 2, 3, \ldots). \tag{9}$$

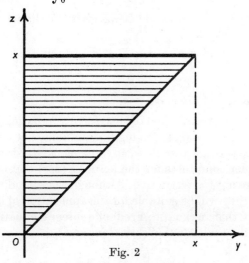

Fig. 2

Moreover, it is easily seen that we also have

$$K_{n+1}(s,y) = \int_0^x K_r(x,z) K_s(z,y)\, dz \quad (r = 1, 2, \ldots, n;\ s = n - r + 1). \tag{10}$$

The proof may be carried out by induction by showing that if the formula is true for $r + s \leqslant n$ and for the pair r_0, s_0 with $r_0 + s_0 = n + 1$ then it is true also for $r = r_0 + 1$, $s = s_0 - 1$ because

$$\int_y^x K_{r_0}(x,z) K_{s_0}(z,y)\, dz = \int_y^x K_{r_0}(x,z)\, dz \int_y^x K(z,u) K_{s_0-1}(u,y)\, du$$

$$= \int_y^x K_{s_0-1}(u,y)\, du \int_u^x K_{r_0}(x,z) K(z,u)\, dz = \int_y^x K_{r_0+1}(x,u) K_{s_0-1}(u,y)\, du.$$

† Note that $K_n(x,y) \equiv 0$ if $y > x$, since $K(x,z) \equiv 0$ for $z > x$.

Thus, it is plausible that we should be led to the solution of (2) by means of the sum (should it exist) of the infinite series defined by (7). But, utilizing (9), we may write (7) in the form

$$\phi_n(x) = f(x) + \int_0^x \left[\sum_{\nu=1}^n \lambda^\nu K_\nu(x, y) \right] f(y)\, dy;$$

hence it is also plausible that the solution of (2) will be given by

$$\phi(x) = f(x) - \lambda \int_0^x H(x, y; \lambda) f(y)\, dy, \tag{11}$$

where $H(x, y; \lambda)$ is the *resolvent kernel* given by the series

$$H(x, y; \lambda) = - \sum_{n=0}^\infty \lambda^n K_{n+1}(x, y). \tag{12}$$

1·4. L_2-Kernels and Functions

As we shall see, the method of successive approximations can be applied to a large number of integral equations, not only those of the Volterra type. Thus, it is advantageous to study its convergence under conditions for the kernel $K(x, y)$ and the function $f(x)$ which are not too restrictive. Although these conditions will be quite weak, they will be suitable for the application of this method (but not only this method) to Fredholm integral equations.

By using the well-known *Schwarz inequality*†

$$\left[\int_a^b f(x) g(x)\, dx \right]^2 \leqslant \int_a^b f^2(x)\, dx \int_a^b g^2(x)\, dx, \tag{1}$$

which will be an important tool in the foregoing theory, we can avoid the customary hypothesis that $K(x, y)$ is continuous (and consequently bounded) by placing it in the L_2-space. Namely, we

---

† This important inequality (which can also be extended to multiple integrals, etc.) can be easily proved by considering the non-negative quadratic form in ξ and η:

$$\int_a^b [\xi f(x) + \eta g(x)]^2\, dx = \xi^2 \int_a^b f^2(x)\, dx + 2\xi\eta \int_a^b f(x) g(x)\, dx + \eta^2 \int_a^b g^2(x)\, dx.$$

The sign of equality holds in (1) if and only if the quotient $f(x)/g(x)$ is constant almost everywhere in (a, b).

In the previous formulae and later we shall often write $f^2(x)$, $g^2(x)$, etc., instead of $[f(x)]^2$, $[g(x)]^2$, etc.

shall suppose that the kernel $K(x, y)$ is *quadratically integrable* in the square $(0 \leqslant x \leqslant h, \; 0 \leqslant y \leqslant h)$, where h is a positive constant, i.e. that the integral

$$\| K \|^2 = \int_0^h \int_0^h K^2(x, y) \, dx \, dy \leqslant N^2 \tag{2}$$

exists (at least in the Lebesgue sense†) and is less than a certain constant N^2. Such a kernel will be called an L_2-*kernel* and $\| K \|$ its *norm*.

Similarly, we shall suppose that the given function $f(x)$ of our integral equations is always an L_2-function, i.e. that its norm $\| f \|$, given by

$$\| f \|^2 = \int_0^h f^2(x) \, dx, \tag{3}$$

exists and is finite.

The fact that K is an L_2-kernel has several important consequences. First of all, Fubini's theorem‡ implies that the functions

$$A(x) = \left[\int_0^h K^2(x, y) \, dy \right]^{\frac{1}{2}}, \quad B(y) = \left[\int_0^h K^2(x, y) \, dx \right]^{\frac{1}{2}} \tag{4}$$

exist almost everywhere for $0 \leqslant x \leqslant h$ and $0 \leqslant y \leqslant h$ respectively, that $A(x)$ and $B(y)$ belong to the class L_2 and finally that

$$\| K \|^2 = \int_0^h A^2(x) \, dx = \int_0^h B^2(y) \, dy. \tag{5}$$

Secondly, if $\Phi(x)$ is any L_2-function in $(0, h)$, then the two functions

$$\psi(x) = \int_0^h K(x, y) \, \Phi(y) \, dy, \quad \chi(y) = \int_0^h K(x, y) \, \Phi(x) \, dx \tag{6}$$

are also L_2-functions. This is an immediate consequence of the Schwarz inequality: moreover, we see that

$$\| \psi \| \leqslant \| K \| \cdot \| \Phi \|, \quad \| \chi \| \leqslant \| K \| \cdot \| \Phi \|. \tag{7}$$

---

† In this book 'integrable' will always signify Lebesgue integrability. If the reader is not familiar with this concept, he may safely assume Riemann integrability in almost all of the subsequent discussion. On this subject see the useful observations of Hardy, Littlewood and Pòlya in the forword to their *Inequalities* [15].

‡ See, for instance, S. Saks, *Théorie de l'intégrale*, Warsaw, 1933, p. 89, or the book of C. Goffman cited below (p. 51), p. 253.

In the same way it is easy to show that the *composition* of two L_2-kernels $K(x,y)$ and $H(x,y)$, i.e. the formation of the two new kernels†

$$G_1(x,y) = \int_0^h K(x,z) H(z,y) \, dz, \quad G_2(x,y) = \int_0^h H(x,z) K(z,y) \, dz,$$
(8)

yields two new L_2-kernels such that

$$\| G_1 \| \leqslant \| K \| . \| H \|, \quad \| G_2 \| \leqslant \| K \| . \| H \|,$$
(9)

and so on. In particular, the last formulae give us useful bounds for the norms of the iterated kernels:

$$\| K_n \| \leqslant \| K \|^n.$$
(10)

This can be immediately proved by induction using (1·3-9).

Finally, we note that it is sometimes suitable to state the stronger condition that the functions (4) not only belong to the class L_2 but are both *bounded*. In such a case we shall say that our kernel belongs to the class L_2^*.

1·5. Solution of Volterra Integral Equations of the Second Kind

The main purpose of this section is the 'legalization' of the results of § 1·3 by proving the following basic theorem:

The Volterra integral equation of the second kind

$$\phi(x) - \lambda \int_0^x K(x,y) \phi(y) \, dy = f(x) \quad (0 \leqslant x \leqslant h),$$
(1)

where the kernel $K(x,y)$ and the function $f(x)$ belong to the class L_2, has one and essentially only one‡ solution in the same class L_2. This solution is given by the formula

$$\phi(x) = f(x) - \lambda \int_0^x H(x,y; \lambda) f(y) \, dy,$$
(2)

---

† Volterra would call this 'composition *of the second kind*'. For composition of the first kind, see § 1·9.

‡ This means that we ignore solutions which are *zero-functions*, i.e. solutions vanishing almost everywhere. (The term 'almost everywhere' is commonly used to designate: 'except for a set whose measure is zero.')

where the 'resolvent kernel' $H(x, y; \lambda)$ is given by the series of iterated kernels

$$-H(x, y; \lambda) = \sum_{\nu=0}^{\infty} \lambda^{\nu} K_{\nu+1}(x, y). \tag{3}$$

Series (3) *converges almost everywhere. The resolvent kernel satisfies the integral equations*

$$K(x, y) + H(x, y; \lambda) = \lambda \int_y^x K(x, z) H(z, y; \lambda) \, dz$$

$$= \lambda \int_y^x H(x, z; \lambda) K(z, y) \, dz. \tag{4}$$

For the proof we shall first show that in this case we can obtain a better bound than (1·4-10) for the norms of the iterated kernels, without changing our basic hypothesis about the given kernel.

In fact, with the help of the Schwarz inequality, we first find

$$K_2^2(x, y) = \left[\int_y^x K(x, z) K(z, y) \, dz \right]^2 \leqslant \int_y^x K^2(x, z) \, dz \int_y^x K^2(z, y) \, dz$$

$$\leqslant \int_0^h K^2(x, z) \, dz \int_0^h K^2(z, y) \, dz = A^2(x) B^2(y),$$

and successively

$$K_3^2(x, y) \leqslant \int_y^x K^2(x, z) \, dz \int_y^x K_2^2(z, y) \, dz$$

$$\leqslant \int_0^h K^2(x, z) \, dz \int_y^x A^2(z) B^2(y) \, dz$$

$$= A^2(x) B^2(y) \int_y^x A^2(z) \, dz,$$

$$K_4^2(x, y) \leqslant \int_y^x K^2(x, z) \, dz \int_y^x K_3^2(z, y) \, dz$$

$$\leqslant \int_0^h K^2(x, z) \, dz \int_y^x A^2(z) B^2(y) \, dz \int_y^z A^2(u) \, du$$

$$= A^2(x) B^2(y) \int_y^x A^2(z) \, dz \int_y^z A^2(u) \, du,$$

....

In general, we can write

$$K_{n+2}^2(x, y) \leqslant A^2(x) B^2(y) F_n(x, y) \quad (n = 1, 2, 3, \ldots), \tag{5}$$

where $\quad F_1(x, y) = \int_y^x A^2(z)\, dz, \quad F_2(x, y) = \int_y^x A^2(z)\, F_1(z, y)\, dz, \ldots$

or, generally,

$$F_n(x, y) = \int_y^x A^2(z)\, F_{n-1}(z, y)\, dz \quad (n = 2, 3, \ldots). \tag{6}$$

Now we state that

$$F_n(x, y) = \frac{1}{n!}\, F_1^n(x, y) \quad (n = 1, 2, 3, \ldots). \tag{7}$$

This formula is obviously valid for $n = 1$. If it is assumed true for $n - 1$, it also remains valid for n, since it follows from (6) that

$$\begin{aligned}
F_n(x, y) &= \frac{1}{(n-1)!} \int_y^x A^2(z)\, F_1^{n-1}(z, y)\, dz \\
&= \frac{1}{(n-1)!} \int_y^x F_1^{n-1}(z, y)\, \frac{\partial F_1(z, y)}{\partial z}\, dz \\
&= \frac{1}{(n-1)!} \left[\frac{1}{n} F_1^n(z, y) \right]_{z=y}^{z=x} = \frac{1}{n!}\, F_1^n(x, y).
\end{aligned}$$

On the other hand, from (1·4-2) it follows that

$$0 \leqslant F_1(x, y) \leqslant \int_0^h A^2(z)\, dz \leqslant N^2;$$

hence $\qquad\qquad 0 \leqslant F_n(x, y) \leqslant \frac{1}{n!}\, N^{2n},$

and by substituting into (5) we obtain

$$|K_{n+2}(x, y)| \leqslant A(x)\, B(y)\, \frac{N^n}{\sqrt{n!}} \quad (n = 0, 1, 2, \ldots).$$

Neglecting the first term, this shows that the infinite series (1·3-12) which gives the resolvent kernel H, has the *majorant*

$$M(x, y) = |\lambda|\, A(x)\, B(y) \sum_{n=0}^{\infty} \frac{(N|\lambda|)^n}{\sqrt{n!}},$$

where the last series *always* converges because the power series

$$\sum_{n=0}^{\infty} \frac{z^n}{\sqrt{n!}}$$

has an infinite radius of convergence.† This is not sufficient to insure that series (1·3-12) be (uniformly and absolutely) convergent everywhere, but it is sufficient to insure its (absolute) convergence *almost* everywhere, because the functions $A(x)$ and $B(y)$ may become infinite on a subset of $(0, h)$ of measure zero. Despite this fact, a fundamental theorem of Lebesgue allows the integration of the series term by term, because the majorant $M(x, y)$ is obviously an L_2-function.

In such a case, for the sake of brevity, we will say that the series is *almost uniformly* convergent.‡

Among other things, it follows that term by term integration can be used to evaluate

$$\int_y^x K(x, z) H(z, y; \lambda) \, dz, \quad \int_y^x H(x, z; \lambda) K(z, y) \, dz.$$

Thus, recalling that the operation which gives us the successive iterated kernels obeys the *associative* law§, as shown by (1·3-10), i.e. remembering that

$$K_n(x, y) = \int_y^x K_h(x, z) K_{n-h}(z, y) \, dz \quad (h = 1, 2, \dots, n-1), \qquad (8)$$

we obtain the basic equations (4) for the resolvent kernel. The interchanges of order of integration which occur in the proof of (8) (as well as those which occur below) are obviously allowed under our hypothesis (the fact that K and hence K_n and H belong to the L_2-class) by virtue of Fubini's theorem.

With the help of equations (4), it is easy to prove that the function ϕ given by formula (2) satisfies equation (1). Indeed, this function

$$\phi_0(x) = f(x) - \lambda \int_0^x H(x, y; \lambda) f(y) \, dy, \qquad (9)$$

certainly belongs to the class L_2 provided that $f(x)$ belongs to the

---

† In fact, if we put $(n!)^{-\frac{1}{2}} = a_n$, we find that

$$\lim_{n \to \infty} \frac{a_n}{a_{n+1}} = \lim_{n \to \infty} \sqrt{(n+1)} = \infty.$$

‡ See also §2.1.

§ But, in general, not the *commutative* law.

same class, since $\phi_0(x)$ is obtained by adding to $f(x)$ a function of the type (1·4-6). But then we have

$$\phi_0(x) - \lambda \int_0^x K(x,y)\,\phi_0(y)\,dy$$

$$= f(x) - \lambda \int_0^x H(x,y;\lambda)f(y)\,dy - \lambda \int_0^x K(x,y)f(y)\,dy$$

$$+ \lambda^2 \int_0^x K(x,z)\,dz \int_0^z H(z,y;\lambda)f(y)\,dy$$

$$= f(x) - \lambda \int_0^x \left[K(x,y) + H(x,y;\lambda) - \lambda \int_y^x K(x,z)H(z,y;\lambda)\,dz \right] f(y)\,dz$$

$$= f(x).$$

Moreover, we can show that—neglecting zero-functions—the function (9) is the *only solution* of class L_2 of the given equation, or, in other words, that any solution of the *homogeneous* Volterra integral equations of the second kind:

$$\phi(x) - \lambda \int_0^x K(x,y)\,\phi(y)\,dy = 0 \tag{10}$$

in L_2-space is necessarily a *zero-function*. For this we observe that if, for brevity, we call ν the *norm* of the function $\phi(x)$ in the basic interval $(0,h)$

$$\nu^2 = \int_0^h \phi^2(x)\,dx,$$

then from (10), with the help of the Schwarz inequality, it follows that

$$\phi^2(x) \leqslant |\lambda|^2 \int_0^x K^2(x,y)\,dy \int_0^x \phi^2(y)\,dy \leqslant |\lambda|^2 A^2(x)\,\nu^2,$$

and successively

$$\phi^2(x) \leqslant |\lambda|^4 \nu^2 \int_0^x K^2(x,y)\,dy \int_0^x A^2(y)\,dy = |\lambda|^4 \nu^2 A^2(x) \int_0^x A^2(y)\,dy,$$

$$\phi^2(x) \leqslant |\lambda|^6 \nu^2 \int_0^x K^2(x,y)\,dy \int_0^x A^2(y)\,dy \int_0^y A^2(z)\,dz$$

$$= |\lambda|^6 \nu^2 A^2(x) \int_0^x A^2(y)\,dy \int_0^y A^2(z)\,dz,$$

By analogy to (7) we have

$$\int_0^x A^2(y_1)\,dy_1 \int_0^{y_1} \dots \int_0^{y_n} A^2(y_n)\,dy_n$$
$$= \frac{1}{n!}\left[\int_0^x A^2(y)\,dy\right]^n \leqslant \frac{1}{n!}N^{2n}; \quad (11)$$

hence, we can write

$$\phi^2(x) \leqslant |\lambda|^2 \nu^2 A^2(x) \frac{(|\lambda|^2 N^2)^n}{n!} \quad (n=0,1,2,\dots),$$

and this shows that $\phi(x)=0$ at any point where $A(x)$ is finite, because the limit of the right-hand side as $n \to \infty$ is obviously zero.[†]

Among other things, this uniqueness theorem shows that equations (4) (the first *or* the second) are *characteristic* for the *resolvent kernel* in the space L_2. For, if a certain L_2-function $H(x,y;\lambda)$ satisfies either equation, it is necessarily the resolvent kernel corresponding to the kernel $K(x,y)$.

1·6. Volterra Equations of the First Kind

As already indicated, the relation between Volterra integral equations of the first and second kind is simpler than that for Fredholm equations.

Thus, if in an equation of the first kind

$$\int_0^x K(x,y)\,\phi(y)\,dy = f(x) \quad (0 \leqslant x \leqslant h), \quad (1)$$

the '*diagonal*' $K(x,x)$ vanishes nowhere in the basic interval $(0,h)$, and if the derivatives

$$\frac{df(x)}{dx} \equiv f'(x), \quad \frac{\partial K}{\partial x} \equiv K'_x(x,y), \quad \frac{\partial K}{\partial y} \equiv K'_y(x,y) \quad (2)$$

exist and are continuous[‡], the equation can be reduced to one of the second kind in two ways.

[†] Hence the exceptional subset where $\phi(x)$ may be different from *zero* is necessarily a subset of the set, where $A(x) = \infty$. Moreover—if equation (10) is satisfied *everywhere* in $(0,h)$—from the fact that $\phi(x)$ is a zero-function, it follows that $\phi(x)$ must be *identically zero*.

[‡] We state these obviously excessive conditions for the sake of simplicity.

The first and simpler way is to differentiate both sides of (1) with respect to x. We obtain the equation

$$K(x,x)\,\phi(x) + \int_0^x K_x'(x,y)\,\phi(y)\,dy = f'(x), \tag{3}$$

that is,

$$\phi(x) + \int_0^x \frac{K_x'(x,y)}{K(x,x)}\,\phi(y)\,dy = \frac{f'(x)}{K(x,x)} \tag{4}$$

and the reduction is accomplished.

The second way utilizes integration by parts; if we set

$$\int_0^x \phi(y)\,dy = \Phi(x), \tag{5}$$

we obtain the equation

$$f(x) = [K(x,y)\,\Phi(y)]_{y=0}^{y=x} - \int_0^x K_y'(x,y)\,\Phi(y)\,dy, \tag{6}$$

that is,

$$\Phi(x) - \int_0^x \frac{K_y'(x,y)}{K(x,x)}\,\Phi(y)\,dy = \frac{f(x)}{K(x,x)}. \tag{7}$$

For this second process it is apparently not necessary for $f(x)$ to be differentiable. However, the function $\phi(x)$ must finally be calculated by differentiating the function $\Phi(x)$ given by the formula

$$\Phi(x) = \frac{f(x)}{K(x,x)} - \int_0^x H^*(x,y;\,1)\,\frac{f(y)}{K(y,y)}\,dy, \tag{8}$$

where $H^*(x,y;\,\lambda)$ is the resolvent kernel corresponding to $K_y'(x,y)/K(x,x)$. To do this, $f(x)$ must be differentiable.

The most important of the previous conditions is the one concerning the diagonal $K(x,x)$; for, if $K(x,x)$ vanishes at some points of the basic interval $(0,h)$, for instance, at $x=0$, then equations (3) and (6) (which can always be formulated) have a peculiar character, essentially different from that of equations of the second kind. These equations have been called by Picard (who in reality considered equations of the Fredholm type) equations of the *third kind*.†

From another point of view, the vanishing of $K(x,x)$ is a complication similar to the vanishing of the first coefficient of an ordinary linear differential equation as we shall see in § 1·8. This singular case was studied by Lalesco [29] and others from just this viewpoint.

---

† However, if $K(x,x)$ vanishes *identically*, it is sometimes possible to obtain an equation of the second kind by repeating the previous transformations.

1·7. An Example

As an example of the Volterra equation of the second kind we consider

$$\phi(x) - \lambda \int_0^x e^{x-y} \phi(y)\, dy = f(x), \tag{1}$$

i.e. the case in which

$$K(x, y) = e^{x-y} = e^x e^{-y}. \tag{2}$$

By a simple calculation, we have

$$K_2(x, y) = \int_y^x e^{x-z} e^{z-y}\, dz = e^{x-y} \int_y^x dz = e^{x-y}(x-y),$$

$$K_3(x, y) = \int_y^x e^{x-z} e^{z-y}(z-y)\, dz = e^{x-y} \int_y^x (z-y)\, dz = e^{x-y} \frac{(x-y)^2}{2},$$

$$K_4(x, y) = \frac{1}{2} \int_y^x e^{x-z} e^{z-y}(z-y)^2\, dz = \frac{1}{2} e^{x-y} \int_y^x (z-y)^2\, dz = e^{x-y} \frac{(x-y)^3}{3!},$$

$$\dots\dots\dots\dots\dots\dots\dots\dots\dots\dots\dots\dots\dots\dots\dots\dots\dots\dots\dots\dots,$$

or, in general,

$$K_{n+1}(x, y) = e^{x-y} \frac{(x-y)^n}{n!} \qquad (n = 0, 1, 2, \dots).$$

Consequently, from (1·3-12), we have

$$H(x, y; \lambda) = - \sum_{n=0}^{\infty} \lambda^n K_{n+1}(x, y) = - e^{x-y} \sum_{n=0}^{\infty} \frac{[\lambda(x-y)]^n}{n!} = - e^{(1+\lambda)(x-y)},$$

and from (1·3-11), we obtain the simple solution

$$\phi(x) = f(x) + \lambda \int_0^x e^{(1+\lambda)(x-y)} f(y)\, dy. \tag{3}$$

The reason for this amazingly simple result lies in the special nature of the kernel (2) which is a function only of the difference $x - y$, and is simultaneously of the type

$$K(x, y) = A(x)/A(y). \tag{4}$$

Kernels of the type $K(x, y) = k(x - y)$ will be discussed later (§ 1·9), but it is easy to see at once that, if (4) holds, then Volterra's equation of the second kind (or of the first kind) can be reduced to a first-order linear differential equation which is immediately integrable: by putting

$$\phi(x)/A(x) = \phi_1(x), \quad f(x)/A(x) = f_1(x),$$

the equation assumes the form

$$\phi_1(x) - \lambda \int_0^x \phi_1(y)\, dy = f_1(x),$$

or

$$\frac{d\phi_2(x)}{dx} - \lambda \phi_2(x) = f_1(x), \quad \text{where} \quad \phi_2(x) = \int_0^x \phi_1(y)\, dy.$$

Then, observing that $\phi_2(0) = 0$, we have

$$\phi_2(x) = e^{\lambda x} \int_0^x e^{-\lambda y} f_1(y)\, dy$$

and

$$\phi_1(x) = \frac{d\phi_2(x)}{dx} = f_1(x) + \lambda \int_0^x e^{\lambda(x-y)} f_1(y)\, dy,$$

$$\phi(x) = f(x) + \lambda \int_0^x e^{\lambda(x-y)} K(x,y) f(y)\, dy. \tag{5}$$

Naturally, this formula can also be obtained by the previous method of successive approximations.

1·8. Volterra Integral Equations and Linear Differential Equations

The example of the previous section indicates a fundamental relationship between Volterra integral equations and ordinary linear differential equations. Actually, the solution of any differential equation of the type

$$\frac{d^n u}{dx^n} + a_1(x) \frac{d^{n-1} u}{dx^{n-1}} + \ldots + a_n(x)\, u = F(x) \tag{1}$$

with continuous coefficients, together with the initial conditions

$$u(0) = c_0, \quad u'(0) = c_1, \quad \ldots, \quad u^{(n-1)}(0) = c_{n-1}, \tag{2}$$

can be reduced to the solution of a certain Volterra integral equation of the second kind

$$\phi(x) + \int_0^x K(x,y)\, \phi(y)\, dy = f(x). \tag{3}$$

In order to accomplish this, we set

$$\mathscr{D}^n u \equiv \frac{d^n u}{dx^n} = \phi(x), \tag{4}$$

and successively

$$D^{-1}\phi = \int_0^x \phi(y)\,dy,$$

$$D^{-2}\phi = D^{-1}(D^{-1}\phi) = \int_0^x (x-y)\,\phi(y)\,dy,$$

$$\dots\dots\dots\dots\dots\dots,$$

$$D^{-n}\phi = D^{-1}(D^{-n+1}\phi) = \frac{1}{(n-1)!}\int_0^x (x-y)^{n-1}\phi(y)\,dy.$$

Taking into account (2), we observe that

$$\left.\begin{aligned}
&\frac{d^{n-1}u}{dx^{n-1}} = c_{n-1} + D^{-1}\phi, \\[2mm]
&\frac{d^{n-2}u}{dx^{n-2}} = c_{n-1}x + c_{n-2} + D^{-2}\phi, \\[2mm]
&\dots\dots\dots\dots\dots\dots\dots\dots\dots\dots\dots\dots, \\[2mm]
&u = c_{n-1}\frac{x^{n-1}}{(n-1)!} + c_{n-2}\frac{x^{n-2}}{(n-2)!} + \dots + c_1 x + c_0 + D^{-n}\phi.
\end{aligned}\right\} \tag{5}$$

Returning to the differential equation (1), we see that it can be written in the form (3) provided we set

$$K(x,y) = \sum_{h=1}^n a_h(x)\frac{(x-y)^{h-1}}{(h-1)!} \tag{6}$$

and

$$f(x) = F(x) - c_{n-1}a_1(x) - (c_{n-1}x + c_{n-2})\,a_2(x) - \dots$$
$$- \left(c_{n-1}\frac{x^{n-1}}{(n-1)!} + \dots + c_1 x + c_0\right)a_n(x). \tag{7}$$

Conversely, solving (3) with K and F given by (6) and (7) and substituting the value obtained for $\phi(x)$ in the last equation of (5), we obtain the (unique) solution of (1) which satisfies the initial conditions (2).

If the leading coefficient in (1) is not unity but $a_0(x)$, equation (3) becomes

$$a_0(x)\,\phi(x) + \int_0^x K(x,y)\,\phi(y)\,dy = f(x), \tag{8}$$

where K and F are still given by (6) and (7) respectively. If $a_0(x) \neq 0$ in the interval considered, nothing is changed; however, if $a_0(x)$ vanishes at some points, we see that an equation of type (8) (which we have already met in the previous section) is equivalent to a *singular* linear differential equation, at least, when $K(x, y)$ is a polynomial in y.

There are still other methods for reducing an ordinary linear differential equation to a Volterra integral equation. Among these, we shall consider the *method of Fubini*† based on a formal use of Lagrange's method of variation of parameters.

For simplicity we shall illustrate the method for a second-order homogeneous equation which we write in the form

$$u'' + p_1(x)\, u' + p_2(x)\, u = A(x)\, u'' + B(x)\, u' + C(x)\, u, \tag{9}$$

splitting the coefficients into two addends. Of course, this can be done in an infinite number of ways, but we choose one which permits us to find explicitly the general solution of the equation

$$u'' + p_1(x)\, u' + p_2(x)\, u = 0 \tag{10}$$

in a simple manner. Moreover, we require that the leading coefficient $1 - A(x)$ of (9) should not vanish in the interval considered.

Let $F_1(x)$ and $F_2(x)$ be two linearly independent solutions of (10), i.e. two solutions whose Wronskian

$$W(x) = \begin{vmatrix} F_1(x) & F_2(x) \\ F_1'(x) & F_2'(x) \end{vmatrix} \tag{11}$$

does not vanish. We attempt to satisfy (9) by a function of the form

$$u(x) = C_1(x)\, F_1(x) + C_2(x)\, F_2(x), \tag{12}$$

where, in the spirit of the method of variation of parameters, $C_1(x)$ and $C_2(x)$ are to be so chosen that

$$C_1'(x)\, F_1(x) + C_2'(x)\, F_2(x) = 0 \tag{13}$$

---

† I like to credit this method to my unforgettable colleague and friend G. Fubini (1879–1943) who first, I believe, used such a device in 1937 to obtain rapidly the classic asymptotic expansions of the Bessel functions. Fubini used, however, a particular case of the method ($A \equiv B \equiv 0$) which is not essentially different from the method of Liouville-Stekloff so often used by G. Szegö in his *Orthogonal Polynomials* [42] and by R. Langer (since 1934). The general form of the method appears perhaps first in my *Equazioni Differenziali* (1st ed. Torino, Einaudi, 1948; 2nd ed. 1953; English ed. in preparation by Blackie and Son, Glasgow).

holds identically. Substituting (12) into (9) and utilizing (13), we find that

$$C_1'F_1' + C_2'F_2' = \frac{AF_1'' + BF_1' + CF_1}{1 - A}C_1 + \frac{AF_2'' + BF_2' + CF_2}{1 - A}C_2. \quad (14)$$

Finally, solving (13) and (14) simultaneously for C_1' and C_2', we obtain

$$C_1' = -(C_1\Phi_1 + C_2\Phi_2)F_2, \quad C_2' = (C_1\Phi_1 + C_2\Phi_2)F_1 \quad (15)$$

with

$$\Phi_h(x) = \frac{AF_h'' + BF_h' + CF_h}{(1 - A)W} \quad (h = 1, 2). \quad (16)$$

The system (15) of differential equations in $C_1(x)$ and $C_2(x)$ is equivalent to the system of integral equations

$$\left.\begin{aligned}
C_1(x) &= \gamma_1 - \int_{x_0}^{x}[C_1(\xi)\,\Phi_1(\xi) + C_2(\xi)\,\Phi_2(\xi)]\,F_2(\xi)\,d\xi, \\
C_2(x) &= \gamma_2 + \int_{x_0}^{x}[C_1(\xi)\,\Phi_1(\xi) + C_2(\xi)\,\Phi_2(\xi)]\,F_1(\xi)\,d\xi,
\end{aligned}\right\} \quad (17)$$

where

$$\gamma_1 = C_1(x_0), \quad \gamma_2 = C_2(x_0)$$

are constants, which may be determined by the requirement that $u(x)$ and $u'(x)$ take on given values at a certain point $x = x_0$.† By setting

$$C_1(x)\,\Phi_1(x) + C_2(x)\,\Phi_2(x) = \psi(x), \quad (18)$$

multiplying the first of the equations (17) by $\Phi_1(x)$ and the second by $\Phi_2(x)$ and adding, we obtain

$$\psi(x) - \int_{x_0}^{x} K(x, \xi)\,\psi(\xi)\,d\xi = f(x), \quad (19)$$

where

$$K(x, \xi) = \begin{vmatrix} F_1(\xi) & F_2(\xi) \\ \Phi_1(x) & \Phi_2(x) \end{vmatrix}, \quad f(x) = \gamma_1\Phi_1(x) + \gamma_2\Phi_2(x). \quad (20)$$

To avoid having the constants γ_1 and γ_2 in the integral equation, we set

$$\psi(x) = \gamma_1\psi_1(x) + \gamma_2\psi_2(x) \quad (21)$$

and obtain the two single integral equations

$$\psi_h(x) - \int_{x_0}^{x} K(x, \xi)\,\psi_h(\xi)\,d\xi = \Phi_h(x) \quad (h = 1, 2). \quad (22)$$

† The equations determining γ_1 and γ_2 are
$$\gamma_1 F_1(x_0) + \gamma_2 F_2(x_0) = u(x_0), \quad \gamma_1 F_1'(x_0) + \gamma_2 F_2'(x_0) = u'(x_0).$$

Finally, the desired solution of (9) in terms of ψ_1 and ψ_2 is easily shown to be

$$u(x) = \gamma_1 U_1(x) + \gamma_2 U_2(x), \tag{23}$$

with $\quad U_h(x) = F_h(x) + \int_{x_0}^{x} \begin{vmatrix} F_1(\xi) & F_2(\xi) \\ F_1(x) & F_2(x) \end{vmatrix} \psi_h(\xi)\, d\xi \quad (h = 1, 2). \tag{24}$

Since under the weak restrictions of (§ 1·4) a Volterra integral equation of the second kind can be solved by an absolutely and *almost uniformly*† convergent series, Fubini's method '*always*' permits us to obtain fundamental series solutions of (9), independently of the order of magnitude of the coefficients A, B, C on the right-hand side. But if, in particular, these coefficients are 'small', the method is especially useful and can be used to obtain several asymptotic representations of solutions of an equation of the type (9).

For instance, if the equation contains a certain parameter λ and we have

$$A = O(\lambda^{-r}), \quad B = O(\lambda^{-r}), \quad C = O(\lambda^{-r}) \quad (\lambda \to \infty, r > 0) \tag{25}$$

uniformly with respect to x, while the other coefficients p and q (and consequently F_1 and F_2) remain bounded, then we have

$$\Phi_h = O(\lambda^{-r}), \quad K(x, \xi) = O(\lambda^{-r}), \quad f(x) = O(\lambda^{-r}).$$

From (22) it follows that

$$\psi_h = O(\lambda^{-r}).$$

Hence, as a consequence of (24), we obtain the simple asymptotic representation

$$U_h(x) = F_h(x) + O(\lambda^{-r}), \tag{26}$$

or more 'sharply',

$$U_h = F_h(x) + \int_{x_0}^{x} \begin{vmatrix} F_1(\xi) & F_2(\xi) \\ F_1(x) & F_2(x) \end{vmatrix} \Phi_h(\xi)\, d\xi + O(\lambda^{-2r}). \tag{27}$$

1·9. Equations of the Faltung Type (Closed Cycle Type)

In the important particular case of a linear differential equation with constant coefficients, the first reduction method of the previous section led us, as (1·8-6) shows, to a Volterra integral equation of

---

† This signifies that the series converges *uniformly* as long as the kernel belongs to the class L_2^*, or that it has the property indicated in §1.5 when the function $A(x)$ is not bounded.

the second kind whose kernel is a polynomial in $x - y$, i.e. to a particular kernel of the type

$$K(x, y) = k(x - y), \tag{1}$$

where $k(t)$ is a certain function of one variable. Such equations

$$\phi(x) - \lambda \int_0^x k(x - y)\, \phi(y)\, dy = f(x), \tag{2}$$

and similar ones of the first kind

$$\int_0^x k(x - y)\, \phi(y)\, dy = f(x), \tag{3}$$

are a very important special class of Volterra integral equations, which Volterra called *equations of the closed cycle* because the operator

$$V_x[\phi(y)] \equiv \int_{-\infty}^x K(x, y)\, \phi(y)\, dy$$

carries any periodic function $\phi(y)$ with arbitrary period T into another periodic function with the same period T, if and only if K is of type (1).[†]

Today they are usually called *equations of the Faltung type* because the operation

$$\phi * \psi \equiv \int_0^x \phi(x - y)\, \psi(y)\, dy \tag{4}$$

is generally called *Faltung* (or *convolution*) of the two functions ϕ and ψ.

The *Faltung* is a particular case of Volterra's *composition of the first kind* (cf. § 1·4)

$$\overset{**}{\Phi\Psi} \equiv \int_y^x \Phi(x, z)\, \Psi(z, y)\, dz, \tag{5}$$

when both functions Φ and Ψ are of type (1).

In fact, if

$$\Phi(x, y) = \phi(x - y), \quad \Psi(x, y) = \psi(x - y),$$

and we put $z = y + t$, then we have

$$\overset{**}{\Phi\Psi} = \int_0^{x-y} \phi(x - y - t)\, \psi(t)\, dt;$$

---

† F. Tricomi, 'Sul "principio del ciclo chiuso" del Volterra', *Atti R. Accad. Sci. Torino*, **76**, 1940–1, pp. 74–82.

that is,

$$\overset{*}{\Phi}\overset{*}{\Psi} = \chi(x-y) \quad \text{with} \quad \chi(u) = \int_0^u \phi(u-t)\,\psi(t)\,dt = \phi * \psi. \quad (6)$$

Like the composition, the convolution is always *associative*

$$\phi * (\psi * \chi) = (\phi * \psi) * \chi, \quad (7)$$

but it is also *commutative*, while Volterra's composition generally is *not* commutative.

In fact, setting $x - y = z$, we obtain

$$\phi * \psi = \int_0^x \phi(z)\,\psi(x-z)\,dz = \psi * \phi. \quad (8)$$

Finally, we notice that it is often convenient to set

$$\phi * \phi = \phi^{*2}, \quad \phi * \phi^{*2} = \phi^{*3}, \quad \dots. \quad (9)$$

Using the Faltung sign, the previous equations (2) and (3) can be written respectively as

$$\phi(x) - \lambda k(x) * \phi(x) = f(x), \quad (2')$$

$$k(x) * \phi(x) = f(x). \quad (3')$$

The main device for dealing with such equations is the *Laplace* transformation

$$\mathscr{L}_x[F(t)] \equiv \int_0^\infty e^{-xt} F(t)\,dt,$$

because, under some restrictions, this operator transforms the convolution into an ordinary product

$$\mathscr{L}_x[F_1(t) * F_2(t)] = \mathscr{L}_x[F_1(t)]\,\mathscr{L}_x[F_2(t)]. \quad (10)$$

In this way, for instance, from equation (3') we obtain

$$\mathscr{L}_x[\phi(t)] = \mathscr{L}_x[f(t)]/\mathscr{L}_x[k(t)], \quad (11)$$

and the resolution of the integral equation is thus reduced to the *inversion* of the Laplace transformation, i.e. to the determination of $F(t)$ when $\mathscr{L}[F(t)]$ is given. But we shall not pursue this method further, since it properly belongs to the theory of the Laplace transformation.†

Aside from this specific method of treatment, it is remarkable

---

† For the Laplace transformation, see, for example, Doetsch [9], Widder [55], Titchmarsh [43].

that all the iterated kernels of (1) as well as the resolvent kernels are of the same type (1). In fact, using the notation of (9), we have

$$K_n(x,y) = k_n(x-y) \quad \text{with} \quad k_n(t) = k^{*n}(t) \quad (n=1,2,\ldots) \quad (12)$$

and $\quad H(x,y;\lambda) = h(x-y,\lambda) \quad \text{with} \quad h(t,\lambda) = -\sum_{n=1}^{\infty} k^{*n}(t)\lambda^{n-1}. \quad (13)$

G. C. Evans† considered the particularly interesting case

$$k(t) = \frac{t^{m-1}}{(m-1)!}, \quad (14)$$

where m is a positive integer, i.e. the case in which $K(x,y) = 1^{*m}$.

By virtue of the formula

$$\frac{t^m}{m!} * \frac{t^n}{n!} = \frac{t^{m+n+1}}{(m+n+1)!},$$

which is an immediate consequence of the fact that

$$\mathscr{L}_x[t^n] = \frac{n!}{x^{n+1}},$$

we obtain $\quad k^{*2}(t) = \dfrac{t^{2m-1}}{(2m-1)!}, \quad k^{*3}(t) = \dfrac{t^{3m-1}}{(3m-1)!}, \ldots,$

and consequently

$$h(t,\lambda) = -\lambda^{(1-m)/m} \sum_{n=1}^{\infty} \frac{(t\lambda^{1/m})^{nm-1}}{(nm-1)!}. \quad (15)$$

This result is remarkable because the sum of the infinite series

$$F_m(z) = \sum_{n=1}^{\infty} \frac{z^{nm-1}}{(nm-1)!} \quad (16)$$

can be expressed in terms of elementary functions.

For this purpose, consider any analytic function regular in the neighborhood of the origin

$$f(z) = a_0 + a_1 z + a_2 z^2 + \ldots,$$

and, for a certain positive integer m, set

$$f_{m,1}(z) = a_0 + a_m z^m + a_{2m} z^{2m} + \ldots,$$
$$f_{m,2}(z) = a_1 z + a_{m+1} z^{m+1} + a_{2m+1} z^{2m+1} + \ldots,$$
$$\cdots\cdots\cdots\cdots\cdots\cdots\cdots\cdots\cdots\cdots\cdots\cdots\cdots\cdots\cdots\cdots\cdots\cdots\cdots\cdots$$
$$f_{m,m}(z) = a_{m-1} z^{m-1} + a_{2m-1} z^{2m-1} + a_{3m-1} z^{3m-1} + \ldots.$$

† See Davis [8], p. 18.

If we denote by $\epsilon_1, \epsilon_2, \ldots, \epsilon_m$ the m complex mth roots of unity, we have

$$\left.\begin{aligned}
f(\epsilon_1 z) &= f_{m,1}(z) + \epsilon_1 f_{m,2}(z) + \epsilon_1^2 f_{m,3}(z) + \ldots + \epsilon_1^{m-1} f_{m,m}(z), \\
f(\epsilon_2 z) &= f_{m,1}(z) + \epsilon_2 f_{m,2}(z) + \epsilon_2^2 f_{m,3}(z) + \ldots + \epsilon_2^{m-1} f_{m,m}(z), \\
&\ldots\ldots\ldots\ldots\ldots\ldots\ldots\ldots\ldots\ldots\ldots\ldots\ldots\ldots\ldots\ldots\ldots, \\
f(\epsilon_m z) &= f_{m,1}(z) + \epsilon_m f_{m,2}(z) + \epsilon_m^2 f_{m,3}(z) + \ldots + \epsilon_m^{m-1} f_{m,m}(z).
\end{aligned}\right\} \quad (17)$$

The determinant of the coefficients of (17) is always different from zero because it is the Cauchy-Vandermonde determinant of the mth roots of unity; hence there exist constants

$$\eta_{rs} \quad (r, s = 1, 2, \ldots, m)$$

depending only on $\epsilon_1, \epsilon_2, \ldots, \epsilon_m$, such that

$$f_{m,r}(z) = \eta_{r1} f(\epsilon_1 z) + \eta_{r2} f(\epsilon_2 z) + \ldots + \eta_{rm} f(\epsilon_m z) \quad (r = 1, 2, \ldots, m) \tag{18}$$

identically.

Applying this result to $F_m(z) = f_{m,m}(z)$, $f(z) = e^z$, we find

$$\left.\begin{aligned}
F_1(z) &= e^z, \quad F_2(z) = \sinh z, \\
F_3(z) &= \tfrac{1}{3} e^z - \tfrac{1}{3} e^{-\frac{1}{2}z} \left[\cos\left(\frac{\sqrt{3}}{2} z\right) + \sqrt{3} \sin\left(\frac{\sqrt{3}}{2} z\right) \right], \\
F_4(z) &= \tfrac{1}{2}(\sinh z - \sin z), \ldots.
\end{aligned}\right\} \quad (19)$$

1·10. Transverse Oscillations of a Bar

The technically important problem of transverse oscillations in a bar can be studied under very general conditions by the method indicated in § 1·1, i.e. by considering the *influence function* $G(x, y)$.

Let us suppose that, in its state of rest, the axis of the bar coincides with the segment $(0, l)$ of the x-axis and that the deflection parallel to the z-axis of a point x at a time t is $z(x, t)$. Then, without more ado, from (1·1-1) and D'Alembert's principle we obtain the *integro-differential* equation

$$z(x, t) = \int_0^l G(x, y) \left[p(y) - \mu(y) \frac{\partial^2 z}{\partial t^2} \right] dy \quad (0 \leqslant x \leqslant l), \tag{1}$$

where $p(y)\, dy$ is the load acting on the portion $(y, y + dy)$ of the bar in the direction Oz, and $\mu(y)\, dy$ is the mass of this portion.

In particular, in the important case of *harmonic vibrations*

$$z(x, t) = Z(x)\, e^{\omega t i} \qquad (2)$$

of an unloaded bar $[p(y) \equiv 0]$, with $l = 1$, we obtain

$$Z(x) - \omega^2 \int_0^1 G(x, y)\, \mu(y)\, Z(y)\, dy = 0, \qquad (3)$$

i.e. *the same homogeneous integral equation as in* § 1·1.

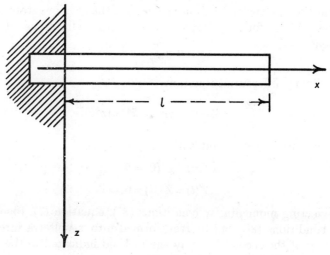

Fig. 3

The fact that the previous equation is the same as that of § 1·1 is very interesting. For instance, this shows the possibility of determining experimentally the critical velocities of a rotating shaft by means of the experimentally simpler harmonic analysis of its transverse oscillations.

Besides, equation (3) shows that our vibration problem belongs properly, as does the problem of § 1·1, to the theory of Fredholm integral equations, and later (§ 3·14) we shall consider equation (3) from that point of view again. But in some cases it is possible to obtain quite precise results, even by means of the more elementary theory of Volterra integral equations.

For instance, this happens in the case of a *uniform* bar

$$\mu(x) - \mu = \text{constant}$$

clamped at the end $x = 0$ and *free* at the end $x = l$. Its transverse oscillations are governed by the partial differential equation†

$$\frac{\partial^4 z}{\partial x^4} + \frac{\mu}{j}\frac{\partial^2 z}{\partial t^2} = 0, \tag{4}$$

where $j = EI$ is the constant *bending rigidity* of the bar, together with

$$z(0, t) = \left[\frac{\partial z}{\partial x}\right]_{x=0} = 0, \quad \left[\frac{\partial^2 z}{\partial x^2}\right]_{x=l} = \left[\frac{\partial^3 z}{\partial x^3}\right]_{x=l} = 0 \tag{5}$$

as conditions on the ends. If use is made of the previous statement (2), we obtain for the transverse harmonic vibrations of the frequency

$$\nu = \frac{\omega}{2\pi} \tag{6}$$

the ordinary differential equation

$$\frac{d^4 Z}{dx^4} - k^4 Z = 0 \quad \text{with} \quad k^4 = 4\pi^2\nu^2\frac{\mu}{j}, \tag{7}$$

together with the end conditions

$$Z(0) = Z'(0) = 0, \tag{8'}$$

$$Z''(l) = Z'''(l) = 0. \tag{8''}$$

Neglecting momentarily conditions (8″), equation (7), together with conditions (8′), can be transformed into a Volterra integral equation of the second kind by the method indicated in the first part of § 1·8. In this way if we put

$$Z''(0) = c_2, \quad Z'''(0) = c_3,$$

we obtain the equation

$$\phi(x) - k^4\int_0^x \frac{(x-y)^3}{3!}\,\phi(y)\,dy = k^4\left(c_2\frac{x^2}{2!} + c_3\frac{x^3}{3!}\right). \tag{9}$$

Hence we obtain an equation with the special kernel (1·9-14) in the case $m = 4$ and consequently from (1·9-15) and (1·9-16) we see that the corresponding resolvent kernel is

$$H(x, y; k^4) = -k^3 F_4[k(x-y)],$$

with $$F_4(z) \equiv f_{4,4}(z) = \frac{z^3}{3!} + \frac{z^7}{7!} + \ldots = \tfrac{1}{2}(\sinh z - \sin z),$$

---

† See, for instance, P. M. Morse [31], p. 114.

and we have

$$\phi(x) = k^4\left(c_2\frac{x^2}{2!} + c_3\frac{x^3}{3!}\right) + k^5 F_4(kx) * \left(c_2\frac{x^2}{2!} + c_3\frac{x^3}{3!}\right).$$

But, using the notations of § 1·9, we find

$$F_4(kx) * \frac{x^2}{2!} = \left(k^3\frac{x^3}{3!} + k^7\frac{x^7}{7!} + \ldots\right) * \frac{x^2}{2!} = k^3\frac{x^6}{6!} + k^7\frac{x^{10}}{10!} + \ldots$$

$$= k^{-3}\left[\frac{(kx)^6}{6!} + \frac{(kx)^{10}}{10!} + \ldots\right] = k^{-3}\left[f_{4,3}(kx) - \frac{(kx)^2}{2!}\right],$$

$$F_4(kx) * \frac{x^3}{3!} = \left(k^3\frac{x^3}{3!} + k^7\frac{x^7}{7!} + \ldots\right) * \frac{x^3}{3!} = k^3\frac{x^7}{7!} + k^7\frac{x^{11}}{11!} + \ldots$$

$$= k^{-4}\left[\frac{(kx)^7}{7!} + \frac{(kx)^{11}}{11!}\right] + \ldots = k^{-4}\left[f_{4,4}(kx) - \frac{(kx)^3}{3!}\right],$$

hence

$$\phi(x) = c_2 k^2 f_{4,3}(kx) + c_3 k f_{4,4}(kx).$$

Even better, since from equations (1·9-17) in the case $m = 4$ it follows that

$$f_{4,3}(z) = \tfrac{1}{2}(\cosh z - \cos z), \quad f_{4,4}(z) = \tfrac{1}{2}(\sinh z - \sin z),$$

we obtain

$$\phi(x) = \tfrac{1}{2}c_2 k^2[\cosh(kx) - \cos(kx)] + \tfrac{1}{2}c_3 k[\sinh(kx) - \sin(kx)]. \quad (10)$$

Now we must remember the neglected conditions (8″) at $x = l$.

Since

$$Z'''(x) = c_3 + D^{-1}\phi(x), \quad Z''(x) = c_2 + c_3 x + D^{-2}\phi(x),$$

or, explicitly,

$$Z'''(x) = \tfrac{1}{2}c_2 k[\sinh(kx) - \sin(kx)] + \tfrac{1}{2}c_3[\cosh(kx) + \cos(kx)],$$

$$Z''(x) = \tfrac{1}{2}c_2[\cosh(kx) + \cos(kx)] + \tfrac{1}{2}c_3 k^{-1}[\sinh(kx) + \sin(kx)],$$

we obtain the two linear equations in the unknowns c_2 and c_3

$$\left.\begin{array}{l} k[\sinh(kl) - \sin(kl)]c_2 + [\cosh(kl) + \cos(kl)]c_3 = 0, \\ [\cosh(kl) + \cos(kl)]c_2 + k^{-1}[\sinh(kl) + \sin(kl)]c_3 = 0. \end{array}\right\} \quad (11)$$

Thus, if the determinant of the coefficients is different from zero, the unique solution of this system is $c_2 = c_3 = 0$, and the corresponding solution of the given equation is the trivial one

$$\phi(x) - Z(x) - 0;$$

but if this determinant vanishes, then there are also non-trivial solutions to our problem. This shows that the only possible harmonic vibrations of our bar are those which correspond to the positive roots of the transcendental equation

$$\sinh^2(kl) - \sin^2(kl) = [\cosh(kl) + \cos(kl)]^2,$$

that is, after simplification,

$$\cosh \xi \cos \xi + 1 = 0, \tag{12}$$

where
$$\xi = kl. \tag{13}$$

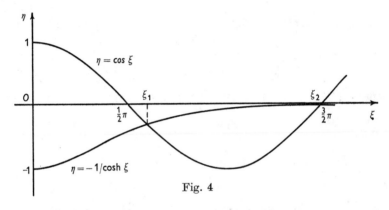

Fig. 4

This transcendental equation† can be solved graphically (Fig. 4) by means of the intersection of the two curves

$$\eta = \cos \xi \quad \text{and} \quad \eta = -1/\cosh \xi.$$

Another way would be to consider the *Gudermannian*

$$\operatorname{gd} \xi = 2 \operatorname{arctg} e^{\xi} - \tfrac{1}{2}\pi$$
$$\scriptstyle (0,\, \frac{1}{2}\pi)$$

(of which there are good numerical tables) which carries the equation into

$$\operatorname{gd} \xi = (2n-1)\pi \pm \xi \quad (n=1, 2, 3, \ldots); \tag{14}$$

a third solution is afforded by the expansion

$$\xi_n = (n-\tfrac{1}{2})\pi + \delta_n - \delta_n^2 + \tfrac{17}{12}\delta_n^3 - \ldots,$$
$$\delta_n = (-1)^{n-1} 2e^{-(n-\frac{1}{2})\pi} \quad (n=1, 2, \ldots), \tag{15}$$

† This equation is also considered in the well-known *Tables of Functions* of Jahnke-Emde (2nd ed., Teubner, 1933, p. 31; ed. of 1948, vol. 1, p. 131), but with some mistakes.

which gives explicitly the successive positive roots $\xi_1, \xi_2, \ldots$ of (12).

Thus we find

$$\xi_1 = 1 \cdot 875106, \quad \xi_2 = 4 \cdot 6941, \quad \ldots$$

By means of (13) and the second equation of (7), we obtain the corresponding natural frequencies of our bar

$$\nu_n = \frac{1}{2\pi} \sqrt{\left(\frac{j}{\mu}\right) \frac{\xi_n^2}{l^2}} \quad (n = 1, 2, 3, \ldots). \tag{16}$$

Contrary to what happens in the case of a vibrating *string*, these frequencies are *not* successive multiples of the first† and are inversely proportional to the square of l, instead of to l itself.

In spite of the thoroughness of the previous result, its interest from the point of view of the theory of integral equations is not great because the key formula (10) for $\phi(x)$ (or the equivalent one for $Z(x)$) can also be obtained directly (and more rapidly) from the linear equation with constant coefficients (7).

From the point of view of the theory of integral equations, it is more interesting that from the integral equation (9) itself we can deduce a good approximation for the first *eigenvalue*

$$k_1^4 = \left(\frac{\xi_1}{l}\right)^4 = \frac{12 \cdot 362}{l^4},$$

because the same device can be used even if the integral equation is not explicitly solvable.

For this, we consider as a first approximation to the unknown function $\phi(x)$ any polynomial of the first degree

$$\phi_0(x) = \alpha + \beta x,$$

and we deduce from the integral equation (9) the second approximation

$$\phi_1(x) = \lambda \left(c_2 \frac{x^2}{2!} + c_3 \frac{x^3}{3!} + \alpha \frac{x^4}{4!} + \beta \frac{x^5}{5!} \right).$$

Thereafter we impose upon both these functions the end conditions

$$c_3 + D^{-1} \phi(x) = 0, \quad c_2 + c_3 x + D^{-2} \phi(x) = 0, \quad (x = 1)$$

† But if n is large, ξ_n is *approximately* equal to all the odd multiples $\frac{1}{2}n\pi$ of $\frac{1}{2}\pi$ (Fig. 4) and ν_n approximately proportional to n^2.

which properly belong to the *exact* function $\phi(x)$ and we obtain four homogeneous linear equations for the four unknowns c_2, c_3, α, β

$$\left.\begin{array}{r}c_3 + \alpha L_1 + \beta L_2 = 0, \\ c_2 + c_3 l + \alpha L_2 + \beta L_3 = 0, \\ c_3 + \lambda(c_2 L_3 + c_3 L_4 + \alpha L_5 + \beta L_6) = 0, \\ c_2 + c_3 l + \lambda(c_2 L_4 + c_3 L_5 + \alpha L_6 + \beta L_7) = 0,\end{array}\right\} \tag{17}$$

where
$$L_n = \frac{l^n}{n!} \quad (n = 1, 2, 3, \ldots, 7).$$

These have a non-trivial solution if and only if

$$\begin{vmatrix} 0 & 1 & L_1 & L_2 \\ 1 & L_1 & L_2 & L_3 \\ L_3 & L_4 + \dfrac{1}{\lambda} & L_5 & L_6 \\ L_4 + \dfrac{1}{\lambda} & L_5 + \dfrac{L_1}{\lambda} & L_6 & L_7 \end{vmatrix} = 0. \tag{18}$$

This is the equation for an approximate evaluation of λ_1. Setting $l = 1$, developing the determinant, and simplifying, we find an equation of the second degree

$$7 \left(\frac{7!}{\lambda}\right)^2 - 408 \left(\frac{7!}{\lambda}\right) + 240 = 0,$$

whose larger root† is 403·84. Consequently,

$$\lambda_1 = 12·480$$

instead of the true value $\lambda_1 = 12·362$, a fairly good approximation!

1·11. Application to the Bessel Functions

We have already noted that Fubini's method, as indicated in the second part of §1·8, was devised in order to obtain easily the classical expansions of Bessel functions for large values of the argument. But the same method can be used, even more advantageously, for the more difficult asymptotic evaluation of Bessel

---

† The other root may be used to obtain a (rough) approximation of λ_2.

functions whose order ν and argument x are almost equal,† i.e. for the study of the behavior as $\nu \to \infty$ of the solutions of the Bessel equation in the neighborhood of the *turning point* $x = \nu$. From a general point of view, in the study of the asymptotic behavior of solutions of the basic differential equation

$$\frac{d}{dx}\left[p(x)\frac{dy}{dx} \right] + [q(x) + \lambda r(x)]\, y = 0, \quad (p(x) > 0), \tag{1}$$

as $\lambda \to \infty$, the *zeros* of the function $r(x)$ are called *turning points* because, depending on whether $r(x) > 0$ or $r(x) < 0$, an integral of (1) either oscillates or is monotonic for large positive values of λ. If $x = x_0$ is a simple zero of the function $r(x)$ and if we set

$$\xi = \int_{x_0}^{x} \frac{dx}{\alpha^2(x)\, p(x)}, \quad y = \alpha(x)\, z, \tag{2}$$

where α is a positive function to be determined later, then under weak conditions on the coefficients our equation is carried into‡

$$\frac{d^2 z}{d\xi^2} + (\lambda \alpha^4 p\rho\xi - Q)\, z = 0,$$

where $\rho = \rho(\xi)$ is a *positive* function in the neighborhood of the turning point $\xi = 0$ and

$$-Q(\xi) = \alpha^3 p(p\alpha'' + p'\alpha' + q\alpha).$$

Hence, if we put $\qquad \alpha = (p\rho)^{-\frac{1}{4}},$

which is possible because $p\rho > 0$, we have the *canonical equation* for the study of turning points

$$\frac{d^2 z}{d\xi^2} + [\lambda\xi - Q(\xi)]\, z = 0. \tag{3}$$

This canonical equation is especially suitable for the application of Fubini's method (§ 1·8) because, with a new change of independent variable

$$\xi = (3\lambda)^{-\frac{1}{3}}\, t, \tag{4}$$

---

† Tricomi, 'Sulle funzioni di Bessel di ordine e argomento pressochè eguali', *Atti Accad. Sci. Torino*, **83**, 1947, pp. 3–20.

‡ See the recent paper of the author 'Equazioni differenziali con punti di transizione (*"turning points"*)', *Rend. Accad. Lincei Roma*, (8) **17**, 1954, II, pp. 137–141.

it can be written $\qquad \dfrac{d^2z}{dt^2} + \tfrac{1}{3}tz = \lambda^{-\frac{2}{3}}Q^*(t)z,$ (5)

with $\qquad\qquad Q^*(t) = 3^{\frac{2}{3}}Q(3^{\frac{1}{3}}\lambda^{-\frac{1}{3}}t).$ (6)

In fact the right side of (5) is $O(\lambda^{-\frac{2}{3}})$ under the weak hypothesis that $Q(\xi)$ is bounded in the neighborhood of $\xi = 0$ and, putting the left side equal to zero, we obtain the classical equation of the Airy function†

$$A_1(t) = \tfrac{1}{3}\pi\sqrt{(\tfrac{1}{3}t)}\{J_{-\frac{1}{3}}[2(\tfrac{1}{3}t)^{\frac{3}{2}}] + J_{\frac{1}{3}}[2(\tfrac{1}{3}t)^{\frac{3}{2}}]\}, \qquad (7)$$

and the similar one

$$A_2(t) = \tfrac{1}{3}\pi\sqrt{t}\{J_{-\frac{1}{3}}[2(\tfrac{1}{3}t)^{\frac{3}{2}}] - J_{\frac{1}{3}}[2(\tfrac{1}{3}t)^{\frac{3}{2}}]\}. \qquad (7')$$

Hence, using only (1·8-26), we have the simple asymptotic representation

$$Z_h = A_h(t) + O(\lambda^{-\frac{2}{3}}), \qquad (8)$$

i.e. $\qquad Z_h(\xi) = A_h[(3\lambda)^{\frac{1}{3}}\xi] + O(\lambda^{-\frac{2}{3}}) \quad (h = 1, 2)$ (9)

of two linear independent solutions Z_1 and Z_2 of the canonical equation (3). This is often sufficient for many purposes.

This explains the character of the 'turning points'.

In fact, depending on whether $\xi > 0$ or $\xi < 0$, the argument $t = (3\lambda)^{\frac{1}{3}}\xi$ of A_h approaches $+\infty$ or $-\infty$ for $\lambda \to \infty$, and the asymptotic behavior of the Airy function $A_h(t)$ as $t \to +\infty$ is completely different from its behavior as $t \to -\infty$, depending on the formulae

$$\left.\begin{aligned} A_h(t) &= \sqrt{\pi}\,(3t)^{-\frac{1}{4}}\cos\left[2\,(\tfrac{1}{3}t)^{\frac{3}{2}} \mp \tfrac{1}{4}\pi\right] + O(t^{-\frac{7}{4}}) \\ &\qquad\qquad (-\text{ for } h=1, \ +\text{ for } h=2), \\ A_1(-t) &= \tfrac{1}{2}\sqrt{\pi}\,(3t)^{-\frac{1}{4}}\exp\{-2(\tfrac{1}{3}t)^{\frac{3}{2}}\}[1 + O(t^{-\frac{3}{2}})], \\ A_2(-t) &= \sqrt{\pi}\,(3t)^{-\frac{1}{4}}\exp\{+2(\tfrac{1}{3}t)^{\frac{3}{2}}\}[1 + O(t^{-\frac{3}{2}})], \end{aligned}\right\} \qquad (10)$$

which can be easily deduced from the classical asymptotic representation of the Bessel functions $J_\nu(x)$ as $x \to \infty$.

Sometimes the reduction of the given equation to a canonical form can be simplified by utilizing the fact that the previous procedure is only slightly changed (if at all) when the right-hand side of (5) also contains terms with z' and z'', but with coefficients of the same order as the coefficient of z.

---

† Watson, *Bessel Functions*, §§ 6·6 and 10·22.

This is the case in the study of Bessel functions of order ν in the neighborhood of the turning point $x = \nu$, because the classical Bessel equation

$$\frac{d^2y}{dx^2} + \frac{1}{x}\frac{dy}{dx} + \left(1 - \frac{\nu^2}{x^2}\right) y = 0, \qquad (11)$$

with the help of the simple substitution

$$x = \nu + (\tfrac{1}{6}\nu)^{\frac{1}{3}} t \qquad (12)$$

may be carried into

$$\frac{d^2y}{dt^2} + \tfrac{1}{3}ty = -\mu \left[(2t + \mu t^2)\frac{d^2y}{dt^2} + (1 + \mu t)\frac{dy}{dt} + \tfrac{1}{6}t^2 y \right], \qquad (13)$$

where $$\mu = 6^{-\frac{1}{3}}\nu^{-\frac{2}{3}}.$$

Consequently, setting $F_1 \equiv A_1(t)$ and $F_2 \equiv A_2(t)$ and

$$K(t, \tau) = A_1(\tau)\,\Phi_2(t) - A_2(\tau)\,\Phi_1(t), \qquad (14)$$

where, according to (1.8–16),

$$\Phi_h(t) = \mu\,\frac{(2t + \mu t^2)[-\tfrac{1}{3}tA_h(t)] + (1 + \mu t)A_h'(t) + \tfrac{1}{6}t^2 A_h(t)}{\tfrac{1}{3}[1 + \mu(2t + \mu t^2)]\pi} \quad (h = 1, 2),$$

$$(15)$$

we obtain from (1·8-23) and (1·8-24)

$$y(t) = \gamma_1 Y_1(t) + \gamma_2 Y_2(t), \qquad (16)$$

with

$$Y_h(t) = A_h(t) + \int_0^t [A_1(\tau)\,A_2(t) - A_2(\tau)\,A_1(t)]\,\phi_h(\tau)\,d\tau \quad (h = 1, 2),$$

$$(17)$$

where ϕ_h is the solution of the Volterra integral equation

$$\phi_h(t) - \int_0^t K(t, \tau)\,\phi_h(\tau)\,d\tau = \Phi_h(t) \quad (h = 1, 2) \qquad (18)$$

and γ_1 and γ_2 are two constants satisfying the equations

$$\gamma_1 A_1(0) + \gamma_2 A_2(0) = y(0), \quad \gamma_1 A_1'(0) + \gamma_2 A_2'(0) = y'(0). \qquad (19)$$

In particular we have, according to (1·8-27),

$$Y_h(t) = A_h(t) + \int_0^t [A_1(t)\,A_2(\tau) - A_2(t)\,A_1(\tau)]\,\Phi_h(\tau)\,d\tau + O(\mu^2). \quad (20)$$

The simple formula (20) already leads us to a better asymptotic approximation formula for the Bessel functions J_ν and N_ν than the more complicated formulae of Nicholson.

For this we first notice that from (15) it follows that

$$\Phi_h(t) = \frac{3}{\pi}[A'_h(t) - \tfrac{1}{2}t^2 A_h(t)]\mu + O(\mu^2),$$

and therefore, considering that

$$\int_0^t \tau^2 A_1(\tau) A_2(\tau)\,d\tau = \tfrac{1}{5}t^3 A_1(t) A_2(t) + \tfrac{3}{5}[A_1(t) - tA'_1(t)]$$
$$\times [A_2(t) - tA'_2(t)] - \tfrac{3}{5}A_1(0) A_2(0),$$

$$\int_0^t \tau^2 A_h(\tau)\,d\tau = \tfrac{1}{5}t^3 A_h(t) + \tfrac{3}{5}[A_h(t) - tA'_h(t)]^2 - \tfrac{3}{5}A_h^2(0),$$

$$\int_0^t A_h(t) A'_h(\tau)\,d\tau = \tfrac{1}{2}A_h^2(t) - \tfrac{1}{2}A_h^2(0),$$

$$\int_0^t A_1(\tau) A'_2(\tau)\,d\tau = \tfrac{1}{2}A_1(t) A_2(t) - \tfrac{1}{6}\pi t - \tfrac{1}{2}A_1(0) A_2(0),$$

$$\int_0^t A'_1(\tau) A_2(\tau)\,d\tau = \tfrac{1}{2}A_1(t) A_2(t) - \tfrac{1}{6}\pi t - \tfrac{1}{2}A_1(0) A_2(0),$$

$$A_2(0) = \sqrt{3}A_1(0) = \tfrac{1}{2}\Gamma(\tfrac{1}{3}), \quad A'_2(0) = -\sqrt{3}A'_1(0) = \tfrac{1}{2}\Gamma(\tfrac{2}{3}),$$

we get

$$Y_h(t) = A_h(t) - \frac{1}{10}\left\{3t^2 A'_h(t) + \left[2t - (-1)^h \frac{\Gamma(\tfrac{1}{3})}{\Gamma(\tfrac{2}{3})}\right]A_h(t)\right.$$
$$\left. + \eta_h \frac{\Gamma(\tfrac{1}{3})}{\Gamma(\tfrac{2}{3})} A_{h+1}(t)\right\}\mu + O(\mu^2), \quad (21)$$

where
$$\eta_1 = +\sqrt{\tfrac{1}{3}}, \quad \eta_2 = -\sqrt{3}, \quad A_3 \equiv A_1.$$

Now we must determine asymptotically the constants γ_1 and γ_2 in the case $y \equiv J_\nu(x)$ as well as in the case $y \equiv N_\nu(x)$ in which we shall write γ_1^* and γ_2^* instead of γ_1 and γ_2 respectively.

This determination is not immediate because it requires a quite sharp asymptotic evaluation of the four constants $J_\nu(\nu)$, $J'_\nu(\nu)$, $N_\nu(\nu)$ and $N'_\nu(\nu)$ as $\nu \to \infty$. This can be done by means of the saddle-point method, whose application is actually very much simplified by the fact that now the argument is *exactly* (not merely approximately) equal to the order ν. The calculations are developed in detail in my paper quoted above. From them we find the formulae†

$$
\left.
\begin{aligned}
J_\nu(\nu) &\sim \frac{1}{2\pi\sqrt{3}}\,(B_0\kappa - B_2\kappa^5 + B_3\kappa^7 - B_5\kappa^{11} + \ldots), \\[4pt]
N_\nu(\nu) &\sim -\frac{1}{2\pi}\,(B_0\kappa + B_2\kappa^5 + B_3\kappa^7 + B_5\kappa^{11} + \ldots), \\[4pt]
J'_\nu(\nu) &\sim \frac{1}{2\pi\sqrt{3}}\,(B'_0\kappa^2 - B'_1\kappa^4 + B'_3\kappa^8 - B'_4\kappa^{10} + \ldots), \\[4pt]
N'_\nu(\nu) &\sim \frac{1}{2\pi}\,(B'_0\kappa^2 + B'_1\kappa^4 + B'_3\kappa^8 + B'_4\kappa^{10} + \ldots),
\end{aligned}
\right\}
\tag{22}
$$

where

$$
\left.
\begin{aligned}
\kappa &= \left(\frac{6}{\nu}\right)^{\frac13}, \qquad
B_n = \sum_{m=0}^{n} \frac{(-6)^m}{m!}\, a_{n-m}^{(m)}\, \Gamma\!\left(\frac{2n+1}{3}+m\right), \\[4pt]
B'_n &= \sum_{m=0}^{n} \frac{(-6)^m}{m!}\, b_{n-m}^{(m)}\, \Gamma\!\left(\frac{2n+2}{3}+m\right), \\[4pt]
\sum_{k=0}^{\infty} a_k^{(m)} z^k &= \left(\frac{1}{5!}+\frac{z}{7!}+\frac{z^2}{9!}+\ldots\right)^m, \\[4pt]
b_k^{(m)} &= a_k^{(m)} + \frac{1}{3!} a_{k-1}^{(m)} + \ldots + \frac{1}{(2k+1)!}\, a_0^{(m)}.
\end{aligned}
\right\}
\tag{23}
$$

With the help of these formulae and of the algebraic linear system (19) we find

$$
\left.
\begin{aligned}
\pi\gamma_1 &= \left(\frac{6}{\nu}\right)^{\frac13} - \frac{1}{10}\frac{\Gamma(\tfrac13)}{\Gamma(\tfrac23)}\frac{1}{\nu} + O(\nu^{-\frac53}), \\[4pt]
\pi\gamma_2 &= \frac{1}{10\sqrt{3}}\frac{\Gamma(\tfrac13)}{\Gamma(\tfrac23)}\frac{1}{\nu} + O(\nu^{-\frac53}), \\[4pt]
\pi\gamma_1^* &= \frac{\sqrt{3}}{10}\frac{\Gamma(\tfrac13)}{\Gamma(\tfrac23)}\frac{1}{\nu} + O(\nu^{-\frac53}), \\[4pt]
\pi\gamma_2^* &= -\left(\frac{6}{\nu}\right)^{\frac13} - \frac{1}{10}\frac{\Gamma(\tfrac13)}{\Gamma(\tfrac23)}\frac{1}{\nu} + O(\nu^{-\frac53});
\end{aligned}
\right\}
\tag{24}
$$

consequently we obtain the formulae

$$
\left.
\begin{aligned}
\pi J_\nu[\nu + (\tfrac16\nu)^{\frac13}t] &= \left(\frac{6}{\nu}\right)^{\frac13} A_1(t) - \frac{1}{10\nu}[3t^2 A'_1(t) + 2t A_1(t)] + O(\nu^{-\frac53}), \\[4pt]
\pi N_\nu[\nu + (\tfrac16\nu)^{\frac13}t] &= -\left(\frac{6}{\nu}\right)^{\frac13} A_2(t) + \frac{1}{10\nu}[3t^2 A'_2(t) + 2t A_2(t)] + O(\nu^{-\frac53}).
\end{aligned}
\right\}
\tag{25}
$$

† The first terms of these formulae coincide numerically with the formulae given (without indication of their origin) at the foot of the numerical tables on pp. 746–7 of Watson's *Bessel Functions*.

These formulae can be utilized to obtain simple asymptotic expressions for the *smallest zeros* $j_{\nu,1}$ and $n_{\nu,1}$ of the functions $J_\nu(x)$ and $N_\nu(x)$ respectively. In fact, I found† (see footnote †, p. 33) that

$$j_{\nu,1} = \nu + 1{\cdot}855757\nu^{\frac{1}{3}} + 1{\cdot}03315\nu^{-\frac{1}{3}} + O(\nu^{-1}),$$
$$n_{\nu,1} = \nu + 0{\cdot}931577\nu^{\frac{1}{3}} + 0{\cdot}26035\nu^{-\frac{1}{3}} + O(\nu^{-1}).$$

1·12. Some Generalizations of the Theory of Volterra Equations

The previous general theory of the Volterra integral equation may be generalized in several different ways. For instance, one may consider kernels of the form

$$K(x,y) = \frac{F(x,y)}{(x-y)^\alpha} \quad (0 < \alpha < 1), \tag{1}$$

where $F(x,y)$ is a continuous (or, at least, a bounded measurable) function.

In spite of the fact that for $\alpha \geqslant \frac{1}{2}$ the square of such a kernel is *not* integrable, the corresponding Volterra integral equation of the second kind can be readily solved. In fact, the successive iterated kernels $K_n(x,y)$ from a certain n on, not only belong to the class L_2, but are *bounded*.

To prove this we use the substitution

$$z = y + (x-y)t,$$

which gives us

$$K_2(x,y) = \int_y^x \frac{F(x,z)\,F(z,y)}{(x-z)^\alpha\,(z-y)^\alpha}\,dz$$
$$= (x-y)^{1-2\alpha} \int_0^1 \frac{F[x,y+(x-y)t]\,F[y+(x-y)t,y]}{t^\alpha\,(1-t)^\alpha}\,dt$$
$$= (x-y)^{1-2\alpha} F_2(x,y), \tag{2}$$

and successively

$$K_3(x,y) = (x-y)^{2-3\alpha} F_3(x,y), \quad K_4(x,y) = (x-y)^{3-4\alpha} F_4(x,y), \ldots,$$

where $F_2, F_3, F_4, \ldots$ are certain bounded functions. Hence K_n, $K_{n+1}, \ldots$ are all bounded as soon as $(n-1)(1-\alpha) > \alpha$.

† Watson (§ 15·81, p. 517) gives only the formulae
$$j_{\nu,1} = \nu + 1{\cdot}855757\nu^{\frac{1}{3}} + O(1), \quad n_{\nu,1} = \nu + 0{\cdot}931577\nu^{\frac{1}{3}} + O(1);$$
the author finds it fascinating that with other methods one can prove that both remainders are $O(\nu^{-\frac{1}{3}})$ instead of $O(1)$.

On the other hand, the equation

$$\phi(x) - \lambda \int_0^x K(x,y)\,\phi(y)\,dy = f(x)$$

can always be reduced to a similar one with kernel K_2, K_3 or K_4. For, by means of a *composition* (cf. § 1·4 and (1·9-5)) of both sides of the equation with $\lambda K(x,y)$, we obtain

$$\phi(x) - \lambda^2 \int_0^x K_2(x,y)\,\phi(y)\,dy = f_2(x) \equiv f(x) + \lambda \int_0^x K(x,y)f(y)\,dy, \quad (3)$$

and then successively

$$\phi(x) - \lambda^3 \int_0^x K_3(x,y)\,\phi(y)\,dy = f_3(x) \equiv f_2(x) + \lambda \int_0^x K(x,y)f_2(y)\,dy,$$

....

A similar transformation is even easier for an integral equation of the first kind, e.g. the important *Abel equation*

$$\int_0^x \frac{\phi(y)}{\sqrt{(x-y)}}\,dy = f(x), \quad (4)$$

where $F \equiv 1$, $\alpha = \tfrac{1}{2}$.

One encounters this equation in the problem of the *tautochrone*: to determine a curve along which a heavy particle, sliding without friction, descends to its lowest point in a constant time, independent of its initial position, or, more generally, such that the time of descent is a given function of its initial position. Abel solved this problem already in 1825, and in essentially the same manner which we shall use; however, he did not realize the general importance of such types of functional equations.

Now, using (4) we obtain from (2)

$$K_2(x,y) = \int_0^1 [t(1-t)]^{-\frac{1}{2}}\,dt = \pi.$$

Hence, the *composition* of (4) with the kernel gives us

$$\pi \int_0^x \phi(y)\,dy = \int_0^x \frac{f(y)}{\sqrt{(x-y)}}\,dy,$$

and, without further ado, we have the amazingly simple solution

$$\phi(x) = \frac{1}{\pi}\frac{d}{dx}\int_0^x \frac{f(y)}{\sqrt{(x-y)}}\,dy. \quad (5)$$

More explicitly: we can integrate by parts, and, provided the function f is differentiable, we obtain

$$\phi(x) = \frac{1}{\pi} \frac{d}{dx} \left[2\sqrt{x} f(0) + 2\int_0^x \sqrt{(x-y)} f'(y) \, dy \right],$$

that is,
$$\phi(x) = \frac{f(0)}{\pi \sqrt{x}} + \frac{1}{\pi} \int_0^x \frac{f'(y)}{\sqrt{(x-y)}} \, dy. \tag{6}$$

In general, the *generalized Abel equation*

$$\int_0^x \frac{\phi(y)}{(x-y)^\alpha} \, dy = f(x) \quad (0 < \alpha < 1) \tag{7}$$

can readily be solved by composition with the kernel $(x-y)^{\alpha-1}$; for

$$\int_y^x \frac{dz}{(x-z)^\alpha (z-y)^{1-\alpha}} = \int_0^1 \frac{dt}{(1-t)^\alpha t^{1-\alpha}} = \Gamma(\alpha)\,\Gamma(1-\alpha) = \frac{\pi}{\sin(\alpha\pi)}$$

gives us†

$$\phi(x) = \frac{\sin(\alpha\pi)}{\pi} \frac{d}{dx} \int_0^x \frac{f(y)}{(x-y)^{1-\alpha}} \, dy = \frac{\sin(\alpha\pi)}{\pi} \left[\frac{f(0)}{x^{1-\alpha}} + \int_0^x \frac{f'(y)}{(x-y)^{1-\alpha}} \, dy \right]. \tag{8}$$

For further types of *singular Volterra equations*, e.g. equations with $(-\infty, x)$ as the limits of integration or equations with non-integrable kernels, see the book of H. T. Davis [8].

We can further generalize the theory of Volterra equations by considering a *system* of equations

$$\phi_r(x) - \lambda \sum_{s=1}^n \int_0^x K_{r,s}(x,y)\,\phi_s(y)\,dy = f_r(x) \quad (r = 1, 2, \ldots, n). \tag{9}$$

One way to deal with such a system is to solve one of these equations with respect to one of the unknown functions and to substitute the solution obtained in the remaining equations. This gives us another system of type (9), but with $n-1$ unknown functions. For instance, if for simplicity we put $\lambda = 1$, and let

† L. Tonelli in *Math. Annalen*, **99**, 1929, pp. 183–99, undertakes an accurate study of Abel's equation with the conclusion, among other things, that the solution (8) is certainly right when the function f is *absolutely continuous*. (We remember that an absolutely continuous function is differentiable almost everywhere.)

$H(x, y)$ be the resolvent kernel corresponding (for $\lambda = 1$) to $K_{1,1}(x, y)$, then from the first equation of (9) we have

$$\phi_1(x) = f_1(x) + \sum_{s=2}^{n} \int_0^x K_{1,s}(x, y)\, \phi_s(y)\, dy - \int_0^x H(x, y) f_1(y)\, dy$$

$$- \sum_{s=2}^{n} \int_0^x \phi_s(y)\, dy \int_y^x H(x, z)\, K_{1,s}(z, y)\, dz;$$

i.e. we obtain an expression of the form

$$\phi_1(x) = F(x) + \sum_{s=2}^{n} \int_0^x A_s(x, y)\, \phi_s(y)\, dy,$$

where $F(x)$ and the $A_s(x, y)$ are known functions. Consequently, by substituting this expression into the 2nd, 3rd, ... and nth equations of the given system, we can reduce it to a similar system with $n - 1$ equations in the unknown functions $\phi_2, \phi_3, ..., \phi_n$.

Another, possibly more advantageous, way to deal with a system of type (9) is to apply directly the method of successive approximations; we can even utilize the experience gathered in § 1·3 to put

$$\phi_r(x) = f_r(x) + \sum_{m=1}^{\infty} \lambda^m \psi_{r,m}(x) \quad (r = 1, 2, ..., n), \tag{10}$$

where the $\psi_{r,m}$ are new unknown functions to be determined. By substituting into (9) and integrating term by term (which may be justified as in § 1·5) we obtain

$$\sum_{m=1}^{\infty} \lambda^m \psi_{r,m}(x) - \sum_{s=1}^{n} \left[\lambda \int_0^x K_{r,s}(x, y) f_s(y)\, dy \right.$$

$$\left. + \sum_{m=1}^{\infty} \lambda^{m+1} \int_0^x K_{r,s}(x, y)\, \psi_{s,m}(y)\, dy \right] = 0.$$

and putting the coefficients of the successive powers of λ equal to zero, we find

$$\psi_{r,1}(x) = \sum_{s=1}^{n} \int_0^x K_{r,s}(x, y) f_s(y)\, dy,$$

$$\psi_{r,2}(x) = \sum_{s=1}^{n} \int_0^x K_{r,s}(x, y)\, \psi_{s,1}(y)\, dy,$$

and, in general,

$$\psi_{r,m}(x) = \sum_{s=1}^{n} \int_0^x K_{r,s}(x, y)\, \psi_{s,m-1}(y)\, dy. \tag{11}$$

If desired, these results may be improved by considering a class of generalized iterated kernels. For instance, the expression for $\psi_{r,\,2}(x)$ can be put into the form

$$\psi_{r,\,2}(x) = \sum_{s=1}^{n} \sum_{t=1}^{n} \int_0^x K_{r,\,s}(x,z)\,dz \int_0^z K_{s,\,t}(z,y) f_t(y)\,dy,$$

that is,

$$\psi_{r,\,2}(x) = \sum_{t=1}^{n} \int_0^x K_{r,\,t}^{(2)}(x,y) f_t(y)\,dy, \tag{12}$$

with

$$K_{r,\,t}^{(2)}(x,y) = \sum_{s=1}^{n} \int_y^x K_{r,\,s}(x,z) K_{s,\,t}(z,y)\,dz. \tag{13}$$

The proof that the series (10) is absolutely and *almost uniformly*† convergent for any λ (provided the kernels $K_{r,\,s}$ belong to the class L_2) is very similar to that given in § 1·4, and consequently need not be repeated here.

1·13. Non-Linear Volterra Equations

In a certain sense the theory of non-linear Volterra integral equations was born before the theory of linear Volterra equations. This is because the classical method of successive approximations for differential equations‡ in its simplest form consists essentially of transforming the differential equation

$$\frac{dy}{dx} = F(x,y), \quad y(x_0) = y_0$$

into the *non-linear Volterra integral equation*

$$y(x) = y_0 + \int_{x_0}^x F[t,y(t)]\,dt,$$

and solving this equation by means of successive approximations

$$y_0(x) \equiv y_0, \quad y_n(x) = y_0 + \int_{x_0}^x F[t,y_{n-1}(t)]\,dt \quad (n=1,2,\ldots).$$

† See p. 13.

‡ Already used explicitly by Liouville (1838) and successively by Caqué (1864), Fuchs (1870), and Peano (1888), but recognized generally as a fundamental method of analysis only after the publications of Picard (1893), who gave it its most general form.

Such a treatment can be extended without difficulty to the more general non-linear Volterra integral equation

$$\phi(x) = f(x) + \int_0^x F[x, y; \phi(y)] \, dy, \tag{1}$$

under much less restrictive conditions on the functions $F(x, y; z)$ and $f(x)$.†

To be precise, it is sufficient to suppose that for any pair z_1, z_2 we can write

$$| F(x, y; z_1) - F(x, y; z_2) | \leqslant a(x, y) \, | z_1 - z_2 |, \tag{2}$$

and further, that we have

$$\left| \int_0^x F[x, y; f(y)] \, dy \right| \leqslant n(x), \tag{2'}$$

where $a(x, y)$ and $n(x)$ are any two L_2-functions, i.e. two functions such that in the entire basic domain $(0 \leqslant y \leqslant x \leqslant h)$ we have

$$\int_0^x n^2(y) \, dy \leqslant N^2, \quad \int_0^h dx \int_0^x a^2(x, y) \, dy \leqslant A^2, \tag{3}$$

where N^2 and A^2 are two positive constants. Even better, putting

$$\int_0^x a^2(x, y) \, dy = \mathscr{A}^2(x), \tag{4}$$

the second condition can be written as

$$\int_0^h \mathscr{A}^2(x) \, dx \leqslant A^2. \tag{5}$$

As in the linear case we attempt to obtain the solution of our equation as the limit of a sequence $\{\phi_n\}$ of functions whose first element is the given function $\phi_0(x) \equiv f(x)$; the other elements are calculated by the recurrence formula

$$\phi_n(x) = f(x) + \int_0^x F[x, y; \phi_{n-1}(y)] \, dy \quad (n = 1, 2, \ldots). \tag{6}$$

First, we have

$$| \phi_1(x) - \phi_0(x) | = \left| \int_0^x F[x, y; f(y)] \, dy \right| \leqslant n(x),$$

† For the study of (1) in the space of the continuous functions, see a recent paper of T. Sato, *Compositio Math.* **11**, 1953, pp. 271–90.

and in general

$$| \phi_{n+1}(x) - \phi_n(x) | \leqslant \int_0^x | F[x, y; \phi_n(y)] - F[x, y; \phi_{n-1}(y)] | \, dy$$

$$\leqslant \int_0^x a(x, y) | \phi_n(y) - \phi_{n-1}(y) | \, dy \quad (n = 1, 2, 3, \ldots)$$

from which, using the Schwarz inequality, it follows that

$$[\phi_{n+1}(x) - \phi_n(x)]^2 \leqslant \int_0^x a^2(x, y) \, dy \int_0^x [\phi_n(y) - \phi_{n-1}(y)]^2 \, dy$$

$$= \mathscr{A}^2(x) \int_0^x [\phi_n(y) - \phi_{n-1}(y)]^2 \, dy.$$

Consequently we have

$$[\phi_1(x) - \phi_0(x)]^2 \leqslant n^2(x),$$

$$[\phi_2(x) - \phi_1(x)]^2 \leqslant \mathscr{A}^2(x) \int_0^x n^2(y) \, dy \leqslant N^2 \mathscr{A}^2(x),$$

$$[\phi_3(x) - \phi_2(x)]^2 \leqslant N^2 \mathscr{A}^2(x) \int_0^x \mathscr{A}^2(y) \, dy,$$

$$[\phi_4(x) - \phi_3(x)]^2 \leqslant N^2 \mathscr{A}^2(x) \int_0^x \mathscr{A}^2(y) \, dy \int_0^y \mathscr{A}^2(z) \, dz,$$

$$\ldots\ldots\ldots\ldots\ldots\ldots\ldots\ldots\ldots\ldots\ldots,$$

and thus, again using (1·5-7), we see that in general

$$[\phi_{n+2}(x) - \phi_{n+1}(x)]^2 \leqslant N^2 \mathscr{A}^2(x) \frac{1}{n!} \left[\int_0^x \mathscr{A}^2(y) \, dy \right]^n \leqslant N^2 \mathscr{A}^2(x) \frac{A^{2n}}{n!},$$

i.e. $| \phi_{n+2}(x) - \phi_{n+1}(x) | \leqslant N \mathscr{A}(x) \dfrac{A^n}{\sqrt{n!}} \quad (n = 0, 1, 2, \ldots).$

But this shows that the infinite series

$$\Phi(x) = \phi_1(x) + [\phi_2(x) - \phi_1(x)] + [\phi_3(x) - \phi_2(x)] + \ldots, \tag{7}$$

whose nth partial sum is $\phi_n(x)$, *converges absolutely* wherever $\mathscr{A}(x)$ is finite, since (neglecting its first term) it admits the majorant

$$N \mathscr{A}(x) \sum_{n=0}^\infty \frac{A^n}{\sqrt{n!}},$$

which is always convergent (§ 1·5); hence we have

$$\lim_{n \to \infty} \phi_n(x) = \Phi(x). \tag{8}$$

and the series converges *almost uniformly* in the sense of p. 13.

Now, in order to prove that the limit function $\Phi(x)$ is a solution of the given equation (1), we set

$$\Phi(x) = \phi_n(x) + R_n(x),\qquad(9)$$

and we observe that $R_n(x)$ is an L_2-function such that

$$\lim_{n\to\infty}\int_0^h R_n^2(x)\,dx = 0,$$

because

$$|R_n(x)| \leqslant N\mathscr{A}(x)\sum_{m=n+1}^{\infty}\frac{A^m}{\sqrt{m!}}.$$

Further we observe that from (6) and (9) it follows that

$$\Phi(x) - f(x) - \int_0^x F[x,y;\ \Phi(y)]\,dy$$
$$= R_n(x) + \int_0^x \{F[x,y;\ \phi_{n-1}(y)] - F[x,y;\ \Phi(y)]\}\,dy,$$

and consequently

$$\left\{\Phi(x) - f(x) - \int_0^x F[x,y;\ \Phi(y)]\,dy\right\}^2$$
$$\leqslant 2R_n^2(x) + 2\left\{\int_0^x a(x,y)\,|R_{n-1}(y)|\,dy\right\}^2.$$

But using the Schwarz inequality

$$\left\{\int_0^x a(x,y)\,|R_{n-1}(y)|\,dy\right\}^2$$
$$\leqslant \mathscr{A}^2(x)\int_0^x R_{n-1}^2(y)\,dy \leqslant \mathscr{A}^2(x)\int_0^h R_{n-1}^2(y)\,dy,$$

hence

$$\int_0^h\left\{\Phi(x) - f(x) - \int_0^x F[x,y;\ \Phi(y)]\,dy\right\}^2 dx$$
$$\leqslant 2\int_0^h R_n^2(x)\,dx + 2A^2\int_0^h R_{n-1}^2(y)\,dy,$$

and, passing to the limit for $n\to\infty$, this shows that the integral on the left is *zero* everywhere $\mathscr{A}(x)$ is finite, i.e. that $\Phi(x)$ satisfies equation (1) almost everywhere.

Moreover, if $f(x)$ belongs to the class L_2, $\Phi(x)$ belongs also to this class, since the functions $\phi_n(x)$ are all L_2-functions and the series (7) converges almost uniformly.

Finally, we shall prove that $\Phi(x)$ is the *unique solution* of the given equation (1) in the space L_2, neglecting almost everywhere vanishing functions.

In fact if there is another solution Φ^* of the class L_2 of the same equation, we have

$$\Phi(x) - \Phi^*(x) = \int_0^x \left\{ F[x, y; \Phi(y)] - F[x, y; \Phi^*(y)] \right\} dy,$$

and, by methods similar to those used above, it follows that

$$[\Phi(x) - \Phi^*(x)]^2 \leqslant \left\{ \int_0^x |F[x, y; \Phi(y)] - F[x, y; \Phi^*(y)]| \, dy \right\}^2$$

$$\leqslant \left\{ \int_0^x a(x, y) |\Phi(y) - \Phi^*(y)| \, dy \right\}^2$$

$$\leqslant \int_0^x a^2(x, y) \, dy \int_0^x [\Phi(y) - \Phi^*(y)]^2 \, dy,$$

i.e. $\qquad [\Phi(x) - \Phi^*(x)]^2 \leqslant \mathscr{A}^2(x) \int_0^x [\Phi(y) - \Phi^*(y)]^2 \, dy. \qquad (10)$

Consequently, if, for the sake of brevity, we put

$$\int_0^h [\Phi(y) - \Phi^*(y)]^2 \, dy = k^2$$

with successive substitutions into (10) we have

$$[\Phi(x) - \Phi^*(x)]^2 \leqslant k^2 \mathscr{A}^2(x),$$

$$[\Phi(x) - \Phi^*(x)]^2 \leqslant k^2 \mathscr{A}^2(x) \int_0^x \mathscr{A}^2(y) \, dy,$$

$$[\Phi(x) - \Phi^*(x)]^2 \leqslant k^2 \mathscr{A}^2(x) \int_0^x \mathscr{A}^2(y) \, dy \int_0^y \mathscr{A}^2(z) \, dz,$$

$$\dotfill,$$

and in general, using (1·5-7) again,

$$\int_0^x [\Phi(x) - \Phi^*(x)]^2 \, dx \leqslant k^2 \frac{1}{n!} \left[\int_0^x \mathscr{A}^2(x) \, dx \right]^n \leqslant k^2 \frac{A^{2n}}{n!}.$$

Passing to the limit for $n \to \infty$ we obtain the statement.

Thus we have proved that *the non-linear Volterra equation of the second kind* (1), *where f and F are L_2-functions satisfying conditions* (2), (2') *and* (3), *has one and essentially only one solution $\Phi(x)$ of the class L_2. This solution is series* (7) *which converges absolutely and* *almost uniformly, and its terms can be calculated using recurrence formula* (6).

To appreciate the importance of this result, we observe that many problems of *non-linear mechanics* lead to the differential equation

$$\frac{d^2y}{dx^2} + \omega^2 y = \mu f\left(x, y, \frac{dy}{dx}\right),$$

where μ *generally*, but not *always*, denotes a small parameter. Now, with the method of Fubini (§ 1·8), this equation can be immediately carried into the *integro-differential* non-linear equation

$$y(x) - \frac{\mu}{\omega} \int_0^x \sin\left[\omega(x-\xi)\right] f[\xi, y(\xi), y'(\xi)]\, d\xi = \gamma_1 \cos(\omega x) + \gamma_2 \sin(\omega x),$$

(11)

where γ_1 and γ_2 are two arbitrary constants.

We could study this equation with the same methods as above; for the sake of brevity, however, we shall merely remark that, *if the function f does not depend on the derivative y'*, then equation (11) is of type (1) and exactly corresponds to the case

$$\left. \begin{array}{l} \lambda = \dfrac{\mu}{\omega}, \\[2mm] F[x, y;\, \phi(y)] = \sin\left[\omega(x-y)\right] f[y, \phi(y)], \\[2mm] f(x) = \gamma_1 \cos(\omega x) + \gamma_2 \sin(\omega x). \end{array} \right\}$$

(12)

For instance, in the case of electrical oscillations of a circuit containing an iron core†

$$\omega^2 = \frac{A}{C}, \quad \mu = -\frac{B}{C}, \quad f(x, y, y') = y^3,$$

where A, B, C are three positive constants, consequently, we have to consider equation (1) for the case

$$F[x, y;\, \phi(y)] = \sin\left[\omega(x-y)\right] [\phi(y)]^3.$$

† N. Minorsky, *Introduction to non-linear mechanics*, Ann Arbor, J. W. Edwards, 1947, p. 192.

CHAPTER II

Fredholm Equations

2·1. Solution by the Method of Successive Approximations: Neumann's Series

The method of successive approximations, which was successfully applied in §§ 1·3–1·5 to Volterra integral equations of the second kind, can be applied even more easily to the basic Fredholm equation of the second kind:†

$$\phi(x) - \lambda \int K(x, y)\, \phi(y)\, dy = f(x). \tag{1}$$

However, the solution obtained in this way may be only *formal*, i.e. if the modulus $|\lambda|$ of the parameter λ is not *small enough*, the infinite series which gives the resolvent kernel is no longer convergent. This method can be used more easily here than in the previous chapter because now all integrations are to be performed between the limits 0 and 1; consequently, the inversion of the order of such integrations is immediate.

As in § 1·3, we set

$$\phi(x) = f(x) + \lambda \psi_1(x) + \lambda^2 \psi_2(x) + \ldots \tag{2}$$

This is called the *Neumann series*. We find immediately that

$$\psi_1(x) = \int K(x, y) f(y)\, dy,$$

$$\psi_2(x) = \int K(x, y)\, \psi_1(y)\, dy = \int K_2(x, y) f(y)\, dy,$$

$$\psi_3(x) = \int K(x, y)\, \psi_2(y)\, dy = \int K_3(x, y) f(y)\, dy,$$

$$\ldots\ldots\ldots\ldots\ldots\ldots\ldots\ldots\ldots\ldots\ldots\ldots\ldots\ldots\ldots,$$

† In this chapter, the integrals frequently have the limits 0 and 1. For this reason, these limits will generally be omitted. We also omit indicating that the independent variable, x, ranges over $(0, 1)$.

where
$$K_2(x,y) = \int K(x,z)\,K(z,y)\,dz,$$
$$K_3(x,y) = \int K(x,z)\,K_2(z,y)\,dz,$$
$$\dots\dots\dots\dots\dots\dots\dots\dots\dots\dots\dots\dots, \tag{3}$$

or, more generally,

$$K_n(x,y) = \int K_h(x,z)\,K_{n-h}(z,y)\,dz$$
$$(n = 2, 3, \dots;\ h = 1, 2, \dots, n-1;\ K_1 \equiv K). \tag{4}$$

The main difference from the Volterra case is that the Neumann series (2) or, what is the same, the series for the resolvent kernel:

$$-H(x,y;\lambda) = K(x,y) + \lambda K_2(x,y) + \lambda^2 K_3(x,y) + \dots, \tag{5}$$

now converges only for *sufficiently small* $|\lambda|$. In other words, although $H(x,y;\lambda)$ is still an analytic function of λ, it is no longer an *entire* function of λ.

We shall now determine a lower bound for the radius of convergence of the power series (5). We observe that *if we preserve the basic hypothesis of* §1·4 *that the kernel* $K(x,y)$ *is an* L_2-*kernel, i.e. supposing that (since now* $h = 1$*):*

$$\|K\|^2 = \iint K^2(x,y)\,dx\,dy = \int A^2(x)\,dx = \int B^2(y)\,dy \leqslant N^2, \tag{6}$$

where
$$A(x) = \left[\int K^2(x,y)\,dy\right]^{\frac{1}{2}}, \quad B(y) = \left[\int K^2(x,y)\,dx\right]^{\frac{1}{2}}, \tag{7}$$

we have successively

$$K_2^2(x,y) = \left[\int K(x,z)\,K(z,y)\,dz\right]^2 \leqslant A^2(x)\,B^2(y),$$

$$K_3^2(x,y) \leqslant \int K^2(x,z)\,dz \int K_2^2(z,y)\,dz$$

$$\leqslant A^2(x)\,B^2(y) \int A^2(z)\,dz \leqslant A^2(x)\,B^2(y)\,N^2,$$

$$K_4^2(x,y) \leqslant \int K^2(x,z)\,dz \int K_3^2(z,y)\,dz$$

$$\leqslant A^2(x)\,B^2(y)\,N^2 \int A^2(z)\,dz \leqslant A^2(x)\,B^2(y)\,N^4,$$

and hence, in general,

$$| K_{n+2}(x, y) | \leqslant A(x) B(y) N^n \quad (n = 0, 1, 2, \ldots). \tag{8}$$

If we neglect the first term, this proves that (5) has the majorant

$$A(x) B(y) | \lambda | \sum_{n=0}^{\infty} (| \lambda | N)^n;$$

this is a *geometric series* with the 'common ratio' $| \lambda | N$; hence it converges for $| \lambda | N < 1$, i.e. for

$$| \lambda | < \| K \|^{-1}. \tag{9}$$

We thus see that, under this condition, the partial sums of (5) have a majorant of the type

$$CA(x) B(y),$$

where C is a constant, i.e. a majorant which is an L_2-function of both x and y. In other words, (5) is an *almost uniformly* convergent series, hence a series which can be integrated term-by-term in either x or y (by Lebesgue's fundamental theorem). By the theorem of Egoroff-Severini,[†] this series becomes uniformly convergent if the points where $A(x) = \infty$ or $B(y) = \infty$ are eliminated from the intervals $0 \leqslant x \leqslant 1$, $0 \leqslant y \leqslant 1$ by means of suitable covering sets of measure less than an arbitrary $\epsilon > 0$.

Furthermore, the previous considerations show that the resolvent kernel $H(x, y; \lambda)$, given by (5), is an analytic function whose singular points are outside (or on the boundary) of the circle (9).

Since term-by-term integration is permitted, we see that by using (4) under condition (9), we have

$$- \int K(x, z) H(z, y; \lambda) dz = - \int H(x, z; \lambda) K(z, y) dz$$

$$= K_2(x, y) + \lambda K_3(x, y) + \ldots$$

$$= \lambda^{-1} [H(x, y; \lambda) + K(x, y)],$$

that is,
$$K(x, y) + H(x, y; \lambda) = \lambda \int K(x, z) H(z, y; \lambda) dz$$
$$= \int H(x, z; \lambda) K(z, y) dz. \left.\right\} \tag{10}$$

† See, for example, C. Goffman, *Real Functions*, Rinehart Co., New York, 1953, p. 187, where an almost uniformly convergent series is said to be *approximately uniformly* convergent.

Even better, considering that all the terms of this double equality are analytic functions of λ, we can assert that the basic equations (10) for the resolvent kernel are valid (almost everywhere) not only in the circle (9), but in the whole domain of existence of the resolvent kernel H in the complex λ-plane.† From this, just as in § 1·4, it follows that if $f(x)$ belongs to the class L_2, then the given equation (1) has at least one solution of the same class L_2; this solution is

$$\phi(x) = f(x) - \lambda \int H(x, y; \lambda) f(y) \, dy \qquad (11)$$

in the domain of existence, $\mathscr{H}$, of H.

Moreover, it is easy to see that the solution (11) is the *unique* L_2-solution of our equation, not only inside the circle $|\lambda| < \| K \|^{-1}$ but also in the whole domain of existence $\mathscr{H}$.

In fact, if at a point $\lambda = \lambda_0$ of $\mathscr{H}$ the *homogeneous equation*

$$\phi(x) - \lambda \int K(x, y) \phi(y) \, dy = 0 \qquad (12)$$

has a certain non-trivial solution $\phi_0(x)$, then with the help of (10) we obtain

$$
\begin{aligned}
\phi_0(x) &= \lambda_0 \int K(x, y) \phi_0(y) \, dy \\
&= -\lambda_0 \int H(x, y; \lambda_0) \phi_0(y) \, dy + \lambda_0^2 \int \phi_0(y) \, dy \int H(x, z; \lambda_0) K(z, y) \, dz \\
&= -\lambda_0 \int H(x, y; \lambda_0) \phi_0(y) \, dy + \lambda_0^2 \int H(x, z; \lambda_0) \, dz \int K(z, y) \phi_0(y) \, dy \\
&= -\lambda_0 \int H(x, y; \lambda_0) \phi_0(y) \, dy + \lambda_0 \int H(x, z; \lambda_0) \phi_0(z) \, dz = 0,
\end{aligned}
$$

and this shows that the function ϕ_0 vanishes almost everywhere.

The following basic theorem is therefore proved:

To each quadratically integrable kernel $K(x, y)$, there corresponds a resolvent kernel $H(x, y; \lambda)$ which is an analytic function of λ, regular at least inside the circle $|\lambda| < \| K \|^{-1}$ and represented there by the

---

† This important *principle of analytic continuation* is based on the fact that the zeros of an analytic function cannot have any accumulation point inside a regular domain of the function. Consequently, two analytic functions $f(z)$ and $g(z)$ which coincide on any *closed* bounded infinite set must necessarily be *identical*, provided this set is contained in the domain of regularity of both functions.

power series (5). *Let the domain of existence of the resolvent kernel in the complex λ-plane be $\mathscr{H}$. Then, if $f(x)$ belongs to the class L_2, the unique, quadratically integrable solution of Fredholm's equation* (1) *valid in $\mathscr{H}$ is given by formula* (11).

The difficulty is that the resolvent kernel is only *potentially* determined by the power series (5). Furthermore, at this moment we know nothing about the nature and distribution of the singular points of $H(x, y; \lambda)$ and nothing about what will happen to the solution of equation (1) if λ coincides with one of these points.

Sometimes it may be useful to write equation (1) in the shortened form

$$\mathscr{F}_x[\phi(y)] = f(x),$$

in which we introduce the *Fredholm linear operator*

$$\mathscr{F}_x[\phi(y)] \equiv \phi(x) - \lambda \int_0^1 K(x, y)\, \phi(y)\, dy. \tag{13}$$

We also introduce the *associated Fredholm operator*

$$\mathscr{F}_x^*[\phi(y)] \equiv \phi(x) - \lambda \int_0^1 K(y, x)\, \phi(y)\, dy. \tag{14}$$

For any pair of functions $\phi(x)$, $\psi(x)$ such that, in the evaluation of the repeated integral

$$\int_0^1 \psi(x)\, dx \int_0^1 K(x, y)\, \phi(y)\, dy,$$

the order of integration can be interchanged, we obviously have

$$\int \mathscr{F}_x[\phi(y)]\, \psi(x)\, dx = \int \mathscr{F}_x^*[\psi(y)]\, \phi(x)\, dx. \tag{15}$$

This formula is often called *Green's formula* for the operator $\mathscr{F}$ because it can be considered as a far-reaching generalization of Green's formula in the theory of linear differential equations.

2·2. An Example

To obtain an idea of the possible answers to the questions posed in the previous section, it is again appropriate to use the kernel of § 1·7 for a Fredholm integral equation, i.e. to consider the integral equation

$$\phi(x) - \lambda \int_0^1 e^{x-y}\, \phi(y)\, dy = f(x). \tag{1}$$

More simply than in § 1·7, we now have

$$K_2(x, y) = \int e^{x-z}\, e^{z-y}\, dz = e^{x-y} \int dz = K(x, y),$$

with the consequence that *all* the iterated kernels K_n coincide with the given kernel K and the series (2·1-5) becomes

$$-H(x, y; \lambda) = K(x, y)\, (1 + \lambda + \lambda^2 + \ldots). \tag{2}$$

Hence we have
$$H(x, y; \lambda) = \frac{e^{x-y}}{\lambda - 1}, \tag{3}$$

and we thus see that the resolvent kernel is actually an analytic function of λ, regular in the whole λ-plane *except at the point* $\lambda = 1$ *which is a simple pole of the kernel.*

Consequently, by virtue of the theorem of the previous section, we see that equation (1) for all $\lambda \neq 1$ has one and only one solution† given by the formula

$$\phi(x) = f(x) - \frac{\lambda}{\lambda - 1}\, e^x \int e^{-y} f(y)\, dy. \tag{4}$$

The same result can be deduced in a more elementary manner by setting

$$\int_0^1 e^{-y}\, \phi(y)\, dy = \xi.$$

The given equation can then be written as

$$\phi(x) = f(x) + \lambda \xi e^x, \tag{5}$$

and consequently, the unknown quantity ξ must satisfy the equation

$$\xi = \int e^{-x}[f(x) + \lambda \xi e^x]\, dx = \int e^{-x} f(x)\, dx + \lambda \xi,$$

that is,
$$(1 - \lambda)\, \xi = \int e^{-x} f(x)\, dx. \tag{6}$$

Therefore, as long as $\lambda \neq 1$, it follows that

$$\xi = \frac{1}{1 - \lambda} \int e^{-x} f(x)\, dx,$$

† Here and later we shall often omit the phrases *of the class L_2* and *neglecting zero-functions* (or similar expressions) as long as there is no danger of confusion.

and, if we substitute into (5), we find (4) again. On the other hand, if $\lambda = 1$, equality (6) shows that the solution of our equation is *generally impossible*, because the given function $f(x)$ usually will *not* satisfy the condition

$$\int e^{-x} f(x)\, dx = 0. \tag{7}$$

But if this condition is satisfied (for instance if $f(x) \equiv 0$), then there exists an *infinite number of solutions*, given by the formula

$$\phi(x) = f(x) + Ce^x, \tag{8}$$

where C denotes an arbitrary constant.

This discussion is applicable to many other cases. We shall see in §§ 2·3–2·4 that in general the singular points of the resolvent kernel, H, are only certain *poles*, $\lambda_1, \lambda_2, \ldots$ whose position in the λ-plane is independent of x and y. Furthermore, for $\lambda = \lambda_1$ or $\lambda = \lambda_2$, etc., equation (2·1-1) in general has no solutions, while the corresponding homogeneous equation (2·1-12) has an infinite number of *non-trivial* solutions. (A *trivial* solution is a *zero-function*, i.e. a solution which vanishes almost everywhere.)

2·3. Fredholm's Equations with Pincherle-Goursat Kernels

The equation of the previous section could be easily treated because its kernel e^{x-y} can be considered as the product of a function e^x of x alone and a function e^{-y} of y alone. Furthermore, we shall see in this section that if a Fredholm integral equation has a kernel which is the sum of n products of functions of x alone by functions of y alone, then it can be reduced to an algebraic system of n linear equations in n unknowns.† Such kernels are usually called Pincherle-Goursat kernels.

More exactly, we say that the kernel $K(x, y)$ is a *Pincherle-Goursat kernel*, or briefly, a *PG-kernel*, if

$$K(x, y) = \sum_{k=1}^{n} X_k(x)\, Y_k(y), \tag{1}$$

where

$$X_1(x), \quad X_2(x), \quad \ldots, \quad X_n(x); \qquad Y_1(x), \quad Y_2(x), \quad \ldots, \quad Y_n(x)$$

---

† It is to be emphasized that the equivalence between the integral equation and the algebraic system will be *exact*; an approximate equivalence can always be stated, for instance, by replacing the kernel K and the given function f by approximating step functions (cf. § 1·2).

are two sets of *linearly independent* L_2-functions† in the basic interval $(0, 1)$.

If we put

$$\int Y_k(x)\, \phi(x)\, dx = \xi_k \quad (k = 1, 2, ..., n), \tag{2}$$

our basic equation

$$\phi(x) - \lambda \int K(x, y)\, \phi(y)\, dy = f(x) \tag{3}$$

is carried into $\qquad \phi(x) = f(x) + \lambda \sum_{k=1}^{n} \xi_k X_k(x). \tag{4}$

This already shows that the difference $\phi(x) - f(x)$ must necessarily coincide with a suitable linear combination of the functions $X_k(x)$.

Moreover, if we multiply (4) by $Y_h(x)$ $(h = 1, 2, ..., n)$ and then integrate between 0 and 1, we have

$$\xi_h = \int Y_h(x) f(x)\, dx + \lambda \sum_{k=1}^{n} \xi_k \int X_k(x)\, Y_h(x)\, dx,$$

that is, $\qquad \xi_h - \lambda \sum_{k=1}^{n} a_{hk} \xi_k = b_h \quad (h = 1, 2, ..., n),$

where $\qquad \int X_k(x)\, Y_h(x)\, dx = a_{hk}, \quad \int Y_h(x) f(x)\, dx = b_h. \tag{5}$

We thus see that the n unknowns $\xi_1, \xi_2, ..., \xi_n$ must satisfy the following algebraic system of n linear equations:

$$\left.\begin{aligned}
(1 - \lambda a_{11})\, \xi_1 - \lambda a_{12} \xi_2 - ... - \lambda a_{1n} \xi_n &= b_1, \\
- \lambda a_{21} \xi_1 + (1 - \lambda a_{22})\, \xi_2 - ... - \lambda a_{2n} \xi_n &= b_2, \\
...., \\
- \lambda a_{n1} \xi_1 - \lambda a_{n2} \xi_2 - ... + (1 - \lambda a_{nn}) \xi_n &= b_n.
\end{aligned}\right\} \tag{6}$$

† The functions $f_1(x)$, $f_2(x)$, ... (even if infinite in number) are called *linearly independent* if the only linear combination,

$$\mu_1 f_1(x) + \mu_2 f_2(x) + ... + \mu_n f_n(x)$$

of n functions which vanishes *almost everywhere* in the basic interval, is the one for which $\mu_1 = \mu_2 = ... = \mu_n = 0$.

Each function $f(x)$ of a linearly independent set is necessarily a *non-zero* function, and consequently its *norm* must always be positive (never zero).

To each solution $\xi_1^{(0)}$, $\xi_2^{(0)}$, ..., $\xi_n^{(0)}$ of this system there corresponds a solution of equation (3) given by (4); conversely, to each solution of the integral equation, which necessarily must be of the form (4), there corresponds a solution of system (6). Furthermore, to two or more *linearly independent* solutions

$$\xi_1^{(0)}, \xi_2^{(0)}, ..., \xi_n^{(0)}; \; \xi_1^{(1)}, \xi_2^{(1)}, ..., \xi_n^{(1)}; \; ...$$

of the system, i.e. to two or more solutions for which we have

$$\mu^{(0)}\xi_h^{(0)} + \mu^{(1)}\xi_h^{(1)} + ... = 0 \quad (h = 1, 2, ..., n)$$

only if $\mu^{(0)} = \mu^{(1)} = ... = 0$, there correspond two or more linearly independent solutions of the integral equation; hence we can speak of a complete *equivalence* between the Fredholm integral equation of the second kind with the *PG*-kernel (1) and the algebraic system (6).

Consequently, the core of the problem is the investigation of the determinant of the coefficients of the system (6):

$$\mathcal{D}(\lambda) = \begin{vmatrix} 1 - \lambda a_{11} & -\lambda a_{12} & ... & -\lambda a_{1n} \\ -\lambda a_{21} & 1 - \lambda a_{22} & ... & -\lambda a_{2n} \\ \hdotsfor{4} \\ -\lambda a_{n1} & -\lambda a_{n2} & ... & 1 - \lambda a_{nn} \end{vmatrix}, \tag{7}$$

which is a polynomial of the nth degree in λ.

If $\mathcal{D}(\lambda) \neq 0$, then system (6) has one and only one solution given by Cramer's formulae

$$\xi_k = \frac{1}{\mathcal{D}(\lambda)}(\mathcal{D}_{1k}b_1 + \mathcal{D}_{2k}b_2 + ... + \mathcal{D}_{nk}b_n) \quad (k = 1, 2, ..., n),$$

where $\mathcal{D}_{h,k}$ denotes the *cofactor* of the (h, k)th element of determinant (7); correspondingly, equation (3) has the unique solution

$$\phi(x) = f(x) + \frac{\lambda}{\mathcal{D}(\lambda)} \sum_{k=1}^{n} (\mathcal{D}_{1k}b_1 + \mathcal{D}_{2k}b_2 + ... + \mathcal{D}_{nk}b_n) X_k(x), \tag{8}$$

while the corresponding *homogeneous* equation

$$\phi(x) - \lambda \int K(x, y)\phi(y)\, dy = 0 \tag{9}$$

has only the *trivial* solution $\phi(x) \equiv 0$.

Furthermore, considering the expression of b_k in (5), solution (8) can also be written as

$$\phi(x) = f(x) + \frac{\lambda}{\mathscr{D}(\lambda)} \int \left\{ \sum_{k=1}^{n} [\mathscr{D}_{1k} Y_1(y) + \mathscr{D}_{2k} Y_2(y) + \dots \right.$$
$$\left. + \mathscr{D}_{nk} Y_n(y)] X_k(x) \right\} f(y)\, dy,$$

but the sum under the integral sign can be considered as the expansion of the negative of a determinant of the $(n+1)$st order†

$$\mathscr{D}(x,y;\lambda) = \begin{vmatrix} 0 & X_1(x) & X_2(x) & \dots & X_n(x) \\ Y_1(y) & 1-\lambda a_{11} & -\lambda a_{12} & \dots & -\lambda a_{1n} \\ Y_2(y) & -\lambda a_{21} & 1-\lambda a_{22} & \dots & -\lambda a_{2n} \\ \dots\dots\dots\dots\dots\dots\dots\dots\dots\dots\dots\dots\dots\dots \\ Y_n(y) & -\lambda a_{n1} & -\lambda a_{n2} & \dots & 1-\lambda a_{nn} \end{vmatrix} ; \quad (10)$$

hence, we can write

$$\phi(x) = f(x) - \frac{\lambda}{\mathscr{D}(\lambda)} \int \mathscr{D}(x,y;\lambda) f(y)\, dy, \qquad (11)$$

with the previous expression of $\mathscr{D}(x,y;\lambda)$.

We thus see that now *the resolvent kernel $H(x,y;\lambda)$ coincides with the quotient of the determinants* (10) *and* (7)

$$H(x,y;\lambda) = \frac{\mathscr{D}(x,y;\lambda)}{\mathscr{D}(\lambda)}, \qquad (12)$$

i.e. with the quotient of two *polynomials of the n-th degree in λ* (the denominator being independent of x and y), and this has important consequences. At this moment, we shall note just one of these: the only possible singular points of $H(x,y;\lambda)$ in the λ-plane are the roots of the equation $\mathscr{D}(\lambda) = 0$, which will be called the *eigenvalues* of our kernel $K(x,y)$.

If $\mathscr{D}(\lambda) = 0$ the non-homogeneous equation (3) has no solution *in general*, because an algebraic linear system with vanishing deter-

---

† In fact if we develop determinant (10) by the elements of the first row and the corresponding minors by the elements of their first column, we find

$$\mathscr{D}(x,y;\lambda) = \sum_{k=1}^{n} (-1)^{k+1} X_k \sum_{h=1}^{n} (-1)^h Y_h \frac{\mathscr{D}_{hk}}{(-1)^{h+k}}$$
$$= -\sum_{k=1}^{n} X_k \sum_{h=1}^{n} \mathscr{D}_{hk} Y_h.$$

minant can only be solved for *certain* values of the quantities on the right-hand side.

Furthermore, from each non-trivial solution $\xi_1^{(0)}, \xi_2^{(0)}, \ldots, \xi_n^{(0)}$ of the homogeneous algebraic system we obtain a non-trivial solution of the *homogeneous* equation (9), which we call an *eigenfnuction* and vice versa.

To be precise, from the theory of algebraic systems of linear equations† we infer that, if λ coincides with a certain eigenvalue λ_0 for which the determinant $\mathscr{D}(\lambda_0)$ has the *characteristic* p‡ $(1 \leqslant p \leqslant n - 1)$, and we put $n - p = r$, then there are ∞^r solutions of the homogeneous system (6). Furthermore, these solutions can be represented by formulae of the type

$$\xi_k = B_{1k}C_1 + B_{2k}C_2 + \ldots + B_{rk}C_r \quad (k = 1, 2, \ldots, n), \tag{13}$$

where $C_1, C_2, \ldots, C_r$ denote r arbitrary constants and

$$\left.\begin{matrix} B_{11}, & B_{12}, & \ldots & B_{1n} \\ \ldots\ldots\ldots\ldots\ldots\ldots\ldots\ldots \\ B_{r1}, & B_{r2}, & \ldots & B_{rn} \end{matrix}\right\} \tag{14}$$

are r arbitrarily fixed but linearly independent§ solutions of the system in question.

This shows that to each eigenvalue λ_0 of *index* $r = n - p \|$ there corresponds a solution of the homogeneous equation (9) of the form

$$\phi_0(x) = C_1 \phi_{01}(x) + C_2 \phi_{02}(x) + \ldots + C_r \phi_{0r}(x), \tag{15}$$

where $C_1, C_2, \ldots, C_r$ are r arbitrary constants and

$$\phi_{01}(x), \phi_{02}(x), \ldots, \phi_{0r}(x)$$

† See Appendix I and Tricomi [45] or O. Schreier and E. Sperner, *Introduction to Modern Algebra and Matrix Theory*, New York, 1951.

‡ The *characteristic* (or *rank*) of a determinant, or a matrix, is the maximal order of its non-vanishing minors. Consequently, if the characteristic is p this means that there is at least one non-vanishing minor of order p, but that the minors of order $p + 1$ or higher (if they exist) are all zero. We shall use the term *characteristic* instead of *rank* to avoid possible mistakes (see below).

§ This means that if $\mu_1 B_{1k} + \mu_2 B_{2k} + \ldots + \mu_r B_{rk} = 0$ $(k = 1, 2, \ldots, n)$, then necessarily $\mu_1 = \mu_2 = \ldots = \mu_r = 0$.

‖ With Lovitt [30], p. 52, we use this name instead of *rank* to avoid confusion with the characteristic of a matrix or determinant, even though rank would correspond better to terminology used in Italian or German.

are r linearly independent functions† which can be expressed in terms of the B_{hk} as follows:

$$\phi_{0h}(x) = \sum_{k=1}^{n} B_{hk} X_k(x) \quad (h = 1, 2, ..., r). \tag{16}$$

Moreover, we can assume that these functions are *normalized*,‡ i.e. that their *norms* are all equal to unity,

$$\int \phi_{0h}^2(x)\, dx = 1 \quad (h = 1, 2, ..., r). \tag{17}$$

All these eigenfunctions are *annihilated* by the Fredholm operator (2·1-14)

$$\mathscr{F}_x[\phi_{0h}(y)] \equiv 0. \tag{18}$$

Using elementary transformations on the determinant (7), we can see§ that the *index* $r = n - p$ *of an eigenvalue is never larger than its multiplicity* m *as a root of the equation* $\mathscr{D}(\lambda) = 0$. *Moreover, in the important case* $a_{hk} = a_{kh}$ *we have*

$$r = m.$$

Another important fact is that to the given kernel (1) and to the *associated* one

$$K(y, x) = \sum_{k=1}^{n} X_k(y)\, Y_k(x), \tag{19}$$

there corresponds the same function $\mathscr{D}(\lambda)$ and consequently *the same eigenvalues*. This is because the interchange of X_k and Y_k carries a_{hk} into a_{kh} and hence only interchanges the rows and columns of determinant (7).

---

† The linear independence of the functions ϕ_{0h} is an immediate consequence of properties of the solutions (14).

‡ A *non-null* function $f(x)$ can always be *normalized* by dividing it by the square root of its norm N^2. For, if

$$\int f^2(x)\, dx = N^2,$$

then obviously

$$\int [f(x)/N]^2\, dx = 1.$$

---

† See, for example, Tricomi [45] or O. Schreier and E. Sperner, *Introduction to Modern Algebra and Matrix Theory*, New York, 1951. These determinant transformations are substantially the same as those which we shall use later (§ 2·5).

However, the eigenfunctions of the associated kernel, i.e. the non-trivial solutions of the *associated homogeneous equation*

$$\psi(x) - \lambda \int K(y, x)\, \psi(y)\, dy = 0, \tag{20}$$

for $\lambda = \lambda_0$ are *not* the previous functions (16) but other ones,

$$\psi_{0h}(x) = \sum_{k=1}^{n} B_{hk}^* Y_k(x) \quad (h = 1, 2, \ldots, r), \tag{21}$$

where
$$\left. \begin{array}{c} B_{11}^*,\ B_{12}^*,\ \ldots,\ B_{1n}^* \\ \cdots\cdots\cdots\cdots\cdots\cdots \\ B_{r1}^*,\ B_{r2}^*,\ \ldots,\ B_{rn}^* \end{array} \right\} \tag{22}$$

are any r linearly independent solutions of the *associated* homogeneous system

$$\left. \begin{array}{l} (1 - \lambda a_{11})\xi_1 - \lambda a_{21}\xi_2 - \ldots - \lambda a_{n1}\xi_n = 0, \\ -\lambda a_{12}\xi_1 + (1 - \lambda a_{22})\xi_2 - \ldots - \lambda a_{n2}\xi_n = 0, \\ \cdots\cdots\cdots\cdots\cdots\cdots\cdots\cdots\cdots\cdots\cdots\cdots \\ -\lambda a_{1n}\xi_1 - \lambda a_{2n}\xi_2 - \ldots + (1 - \lambda a_{nn})\xi_n = 0. \end{array} \right\} \tag{23}$$

Any eigenfunction $\phi_{0h}(x)$ corresponding to the eigenvalue λ_0 and any *associated* eigenfunction $\psi_{1k}(x)$ corresponding to a *different* eigenvalue λ_1 are always *orthogonal* in the basic interval $(0, 1)$.†
In fact we have

$$I = \int \phi_{0h}(x)\, \psi_{1k}(x)\, dx = \lambda_0 \int \psi_{1k}(x)\, dx \int K(x, y)\, \phi_{0h}(y)\, dy$$

$$= \lambda_0 \int \phi_{0h}(y)\, dy \int K(x, y)\, \psi_{1k}(x)\, dx = \frac{\lambda_0}{\lambda_1} \int \phi_{0h}(y)\, \psi_{1k}(y)\, dy = \frac{\lambda_0}{\lambda_1} I,$$

and this equality can be true only if $\lambda_0 = \lambda_1$ or if $I = 0$.

We now return to the *non-homogeneous equation* (3) *for the case* $\mathscr{D}(\lambda) = 0$. We prove that *for* $\lambda = \lambda_0$ *the non-homogeneous equation can be solved if and only if the* r *orthogonality conditions*

$$(f, \psi_{0h}) \equiv \int f(x)\, \psi_{0h}(x)\, dx = 0 \quad (h = 1, 2, \ldots, r) \tag{24}$$

† In general, two functions $g(x)$ and $f(x)$ are called orthogonal over (a, b) if
$$\int_a^b f(x)\, g(x)\, dx = 0.$$

are satisfied. In this case the non-homogeneous equation has ∞^r solutions of the form

$$\phi(x) = \Phi(x) + C_1\phi_{01}(x) + C_2\phi_{02}(x) + \ldots + C_r\phi_{0r}(x), \qquad (25)$$

where $\Phi(x)$ is a suitable linear combination of $X_1(x), X_2(x), \ldots, X_n(x)$.

In fact, conditions (24) are *necessary* because if equation (3) for $\lambda = \lambda_0$ admits a certain solution $\Phi(x)$, then from the equation itself, it follows that

$$\int f(x)\,\psi_{0h}(x)\,dx = \int \Phi(x)\,\psi_{0h}(x)\,dx - \lambda_0\int \psi_{0h}(x)\,dx\int K(x,y)\,\Phi(y)\,dy$$

$$= \int \Phi(x)\,\psi_{0h}(x)\,dx - \lambda_0\int \Phi(y)\,dy\int K(x,y)\,\psi_{0h}(x)\,dx.$$

But, since λ_0 and $\psi_{0h}(x)$ are eigenvalue and corresponding eigenfunction of the *associated* kernel, we have

$$\lambda_0\int K(x,y)\,\psi_{0h}(x)\,dx = \psi_{0h}(y);$$

hence
$$\int f(x)\,\psi_{0h}(x)\,dx = 0.$$

Furthermore, conditions (24) are also *sufficient*, since from them it can be easily deduced that the non-homogeneous system (6), which we shall write briefly as

$$\Xi_1 = b_1, \quad \Xi_2 = b_2, \quad \ldots, \quad \Xi_n = b_n,$$

reduces to only $n - r$ independent equations. Consequently we can now solve it readily (carrying r unknowns on the right-hand side), since the characteristic of the matrix of the coefficients is exactly $p = n - r$.

We can reduce the system for the following reason: Let us multiply the previous equations by B_{h1}^*, B_{h2}^*, ..., B_{hr}^* respectively and add. Bearing in mind equations (23), we have

$$\sum_{k=1}^{n} B_{hk}^* \Xi_k = [(1 - \lambda a_{11})B_{h1}^* - \lambda a_{21}B_{h2}^* - \ldots - \lambda a_{n1}B_{hn}^*]\xi_1$$

$$+ [-\lambda a_{12}B_{h1}^* + (1 - \lambda a_{22})B_{h2}^* - \ldots - \lambda a_{n2}B_{hn}^*]\xi_2$$

$$+ \ldots\ldots\ldots\ldots\ldots\ldots\ldots\ldots\ldots\ldots\ldots\ldots\ldots\ldots$$

$$+ [-\lambda a_{1n}B_{h1}^* - \lambda a_{2n}B_{h2}^* - \ldots + (1 - \lambda a_{nn})B_{hn}^*]\xi_n \equiv 0,$$

while on the other side, in virtue of (24), we also have

$$\sum_{k=1}^{n} B_{hk}^* b_k = \int \left[\sum_{k=1}^{n} B_{hk}^* Y_k(x) \right] f(x)\, dx = \int \psi_{0h}(x) f(x)\, dx = 0.$$

Among other things, form (25) of the solution demonstrates the following obvious fact: the general solution of equation (3) when $\mathscr{D}(\lambda) = 0$ can be considered as the sum of any particular solution $\Phi(x)$ and of the general solution (15) of the homogeneous equation.

Thus we have proved *for PG-kernels* the following basic *Fredholm theorem*, which will be extended to general kernels in the next section:

Fredholm's integral equation of the second kind

$$\phi(x) - \lambda \int K(x, y)\, \phi(y)\, dy = f(x)$$

has, in general, one and only one solution of the class L_2 given by the formula

$$\phi(x) = f(x) - \lambda \int H(x, y; \lambda) f(y)\, dy,$$

where $H(x, y; \lambda)$ is the resolvent kernel. $H(x, y; \lambda)$ is an analytic function of λ, and if $|\lambda| < \| K \|^{-1}$ it is given by the Neumann series

$$-H(x, y; \lambda) = K(x, y) + \lambda K_2(x, y) + \lambda^2 K_3(x, y) + \dots,$$

where K_2, K_3, ... are the iterated kernels. The only exceptions are the singular points of $H(x, y; \lambda)$ which coincide† with the zeros (called eigenvalues) of an analytic function $\mathscr{D}(\lambda)$ of λ. In the case of a PG-kernel, $\mathscr{D}(\lambda)$ is a polynomial.

If $\lambda = \lambda_0$ is a root of multiplicity $m \geqslant 1$ of the equation $\mathscr{D}(\lambda) = 0$, then the homogeneous equation

$$\phi(x) - \lambda \int K(x, y)\, \phi(y)\, dy = 0$$

has r linearly independent non-trivial solutions, called eigenfunctions, where r, the index of the eigenvalue, satisfies the condition $1 \leqslant r \leqslant m$. The same is true for the associated homogeneous equation

$$\psi(x) - \lambda \int K(y, x)\, \psi(y)\, dy = 0.$$

† If λ_0 is a root of $\mathscr{D}(\lambda)$, then λ_0 is a singular point of $H(x, y; \lambda)$, even if $\mathscr{D}(x, y; \lambda_0) = 0$, because at each eigenvalue the homogeneous equation has at least ∞^1 (if $p = n - 1$) non-trivial solutions, while (see § 2·1) at any regular point of $H(x, y; \lambda)$ this equation has only $\phi(x) \equiv 0$ for a solution.

However, if $\lambda = \lambda_0$, the non-homogeneous equation has solutions (exactly ∞^r solutions) if and only if the given function $f(x)$ is orthogonal to all the eigenfunctions of the associated homogeneous equation.

A very important alternative theorem can immediately be deduced as a corollary:

ALTERNATIVE THEOREM. *If the homogeneous Fredholm integral equation has only the trivial solution, then the corresponding non-homogeneous equation always has one and only one solution. On the contrary, if the homogeneous equation has some non-trivial solutions, then the non-homogeneous integral equation has either no solution or an infinity of solutions, depending on the given function $f(x)$.*

But even this corollary has been proved only for PG-kernels.

Finally, we note that the results of this section remain substantially unchanged if $X_k(x)$, $Y_k(y)$ and $f(x)$ depend (analytically) on the *parameter* λ, i.e. for an equation of the form

$$\phi(x) - \lambda \int \left[\sum_{k=1}^{n} X_k(x, \lambda) Y_k(y, \lambda) \right] \phi(y)\, dy = f(x, \lambda). \tag{26}$$

In this case a_{hk} and b_h become certain *analytic* functions of λ and $\mathscr{D}(\lambda)$ and $\mathscr{D}(x, y; \lambda)$ are no longer *polynomials* in λ, but *analytic functions* of λ of a more general nature. Consequently, it may even occur that there are *no* eigenvalues because a non-algebraic analytic function may have no zeros.

2·4. The Fredholm Theorem for General Kernels

Using a method of E. Schmidt, which was found again and generalized by Picone,† we shall now extend the basic Fredholm theorem to general L_2-kernels, i.e. to kernels satisfying only the few conditions of § 2·1. For this we first observe that each L_2-kernel, $K(x, y)$, can be decomposed (in an infinite number of ways) into the sum of a suitable PG-kernel, $S(x, y)$, and another L_2-kernel, $T(x, y)$, whose norm $\| T \|$ can be made as small as we wish. This will be proved in the next chapter (§ 3·6), naturally, without the use of the Fredholm theorem. To be precise, suppose that all the points of the λ-plane which we want to consider lie inside the circle $| \lambda | = R$ (where R denotes an *arbitrary* positive constant). We put

$$K(x, y) = S(x, y) + T(x, y), \tag{1}$$

† Picone [37], Cap. VII, p. 582 seq.

where

$$S(x,y) = \sum_{k=1}^{n} X_k(x)\, Y_k(y), \quad \|T\|^2 = \iint T^2(x,y)\, dx\, dy < 1/R^2. \quad (2)$$

This gives us condition (2·1-9) and thereby the convergence of the Neumann series for the kernel T.

Consequently, if we call $H_T(x, y; \lambda)$ the resolvent kernel corresponding to $T(x, y)$ since our basic equation can now be written

$$\phi(x) - \lambda \int T(x,y)\, \phi(y)\, dy = F(x),$$

with

$$F(x) = f(x) + \lambda \int S(x,y)\, \phi(y)\, dy;$$

we can replace it with the completely equivalent equation

$$\phi(x) = F(x) - \lambda \int H_T(x, y; \lambda)\, F(y)\, dy,$$

that is, with the equation

$$\phi(x) - \lambda \int \left[S(x,y) - \lambda \int H_T(x, z; \lambda)\, S(z, y)\, dz \right] \phi(y)\, dy$$
$$= f(x) - \lambda \int H_T(x, y; \lambda) f(y)\, dy,$$

which we can write as

$$\phi(x) - \lambda \int \left[\sum_{k=1}^{n} X_k^*(x, \lambda)\, Y_k(y) \right] \phi(y)\, dy = f^*(x, \lambda), \quad (3)$$

where

$$\left.\begin{aligned}
X_k^*(x, \lambda) &= X_k(x) - \lambda \int H_T(x, y; \lambda)\, X_k(y)\, dy, \\
f^*(x, \lambda) &= f(x) - \lambda \int H_T(x, y; \lambda) f(y)\, dy.
\end{aligned}\right\} \quad (4)$$

In this way any Fredholm equation of the second kind with an L_2-kernel can be reduced to a similar one with a PG-kernel, or, more exactly, to an equation of type (2·3-26). But for these equations the basic Fredholm theorem of the previous section is valid; hence this theorem is also valid for every Fredholm integral equation of the second kind with an L_2-kernel.

Moreover, we can immediately see that the *spectrum* (i.e. the set of the eigenvalues) of a Fredholm integral equation of the second kind with an L_2-kernel has no accumulation points, except possibly the point at infinity.

In fact, the eigenvalues inside *any* finite circle $|\lambda| \leqslant R$ coincide with the zeros of an analytic function, $\mathscr{D}_R(\lambda)$: the determinant (7) for equation (3), which is certainly regular for $|\lambda| \leqslant R$.

2·5. The Formulae of Fredholm

In addition to the previous results, we now give (not only for historical reasons) the celebrated *Fredholm formulae* which explicitly represent the resolvent kernel H of an integral equation of the second kind with a bounded† kernel K. This resolvent kernel will be given as the quotient of two *everywhere convergent* power series in λ; this is similar to the case of a *PG*-kernel where the corresponding resolvent kernel is the quotient of two *polynomials* in λ (see (2·3-12)).

The Fredholm formulae can be derived in a quite simple and elegant way, because we can avoid some rather difficult discussions, which would be necessary if the basic Fredholm theorem were not already established, but had to be deduced from these formulae.

The central idea of Fredholm's method is the heuristic construction of a certain function $\mathscr{D}(\lambda)$ which is to play a role similar to that of the function $\mathscr{D}(\lambda)$ of § 2·3. Then one must prove rigorously only that both this function and the product

$$\mathscr{D}(x, y; \lambda) = \mathscr{D}(\lambda) H(x, y; \lambda) \tag{1}$$

are *entire* functions of λ, which will be represented by certain elegant power series expansions.

For the construction of $\mathscr{D}(\lambda)$ we start by assuming that the given bounded kernel $K(x, y)$ can be represented approximately by the *step function* (like those of § 1·2):

$$K(x, y) = K\left(\frac{r}{n}, \frac{s}{n}\right) = k_{rs}, \quad \begin{cases} \dfrac{r-1}{n} < x \leqslant \dfrac{r}{n} & (n = 1, 2, 3, \dots), \\[2mm] \dfrac{s-1}{n} < y \leqslant \dfrac{s}{n} & (r, s = 1, 2, \dots, n). \end{cases}$$

Then (without too many scruples) we shall 'calculate' the limit as $n \to \infty$ of the determinant of the coefficients of the corresponding linear algebraic system (1·2-1′′′).

† Recently S. Mihlin ('On the convergence of Fredholm series', *Comptes Rend. (Doklady) Acad. Sci. USSR* (N.S.), **42**, 1944, pp. 373–6) proved the convergence of Fredholm series even under the hypothesis that the kernel K belongs only to the class L_2.

We first make repeated use of the well-known theorem on determinants with a row (or column) of sums of two addends. Using this theorem, the determinant mentioned above:

$$D_n = \begin{vmatrix} 1-\dfrac{\lambda}{n}k_{11} & 0-\dfrac{\lambda}{n}k_{12} & \ldots & 0-\dfrac{\lambda}{n}k_{1n} \\ 0-\dfrac{\lambda}{n}k_{21} & 1-\dfrac{\lambda}{n}k_{22} & \ldots & 0-\dfrac{\lambda}{n}k_{2n} \\ \cdots\cdots\cdots\cdots\cdots\cdots\cdots\cdots\cdots\cdots\cdots\cdots \\ 0-\dfrac{\lambda}{n}k_{n1} & 0-\dfrac{\lambda}{n}k_{n2} & \ldots & 1-\dfrac{\lambda}{n}k_{nn} \end{vmatrix}$$

may be put into the form

$$D_n = 1 - \frac{\lambda}{n}S_1 + \left(\frac{\lambda}{n}\right)^2 S_2 - \ldots + (-1)^n \left(\frac{\lambda}{n}\right)^n S_n. \tag{2}$$

Here $S_m \, (m = 1, 2, \ldots, n)$ denotes the sum of all the *principal minors* of order m contained in the determinant with the elements k_{rs}, i.e. of all the $\binom{n}{m}$ minors formed with certain m rows of this determinant and with m columns with the *same* index. In other words, we set

$$S_m = \Sigma \begin{vmatrix} k_{r_1 r_1} & k_{r_1 r_2} & \ldots & k_{r_1 r_m} \\ k_{r_2 r_1} & k_{r_2 r_2} & \ldots & k_{r_2 r_m} \\ \cdots\cdots\cdots\cdots\cdots\cdots\cdots \\ k_{r_m r_1} & k_{r_m r_2} & \ldots & k_{r_m r_m} \end{vmatrix},$$

where $r_1, r_2, \ldots, r_m$ denote any combination of m of the n numbers $1, 2, \ldots, n$, arranged in increasing order. Even better, considering that any permutation of the index $r_1, r_2, \ldots, r_m$ leaves unchanged (even in sign) the previous minor,‡ and that each minor, with two or more of the $r_1, r_2, \ldots, r_m$ equal, vanishes, we can allow each index r_h to run (independently) between 1 and n, provided that we divide by $m!$ to compensate for the fact that now each of the earlier terms will be repeated $m!$ times. Hence we can write

$$S_m = \frac{1}{m!} \sum_{r_1=1}^{n} \sum_{r_2=1}^{n} \ldots \sum_{r_m=1}^{n} \begin{vmatrix} K\left(\dfrac{r_1}{n}, \dfrac{r_1}{n}\right) & \ldots & K\left(\dfrac{r_1}{n}, \dfrac{r_m}{n}\right) \\ \cdots\cdots\cdots\cdots\cdots\cdots\cdots\cdots\cdots \\ K\left(\dfrac{r_m}{n}, \dfrac{r_1}{n}\right) & \ldots & K\left(\dfrac{r_m}{n}, \dfrac{r_m}{n}\right) \end{vmatrix},$$

† Because the rows *and* the columns undergo the same permutation.

or, more briefly,

$$S_m = \frac{1}{m!} \sum_{r_1, \ldots, r_m = 1}^{n} K \begin{pmatrix} \dfrac{r_1}{n}, & \dfrac{r_2}{n}, & \ldots, & \dfrac{r_m}{n} \\[2mm] \dfrac{r_1}{n}, & \dfrac{r_2}{n}, & \ldots, & \dfrac{r_m}{n} \end{pmatrix}, \tag{3}$$

where†

$$K \begin{pmatrix} x_1, x_2, \ldots, x_m \\ y_1, y_2, \ldots, y_m \end{pmatrix} = \begin{vmatrix} K(x_1, y_1) & K(x_1, y_2) & \ldots & K(x_1, y_m) \\ K(x_2, y_1) & K(x_2, y_2) & \ldots & K(x_2, y_m) \\ \ldots\ldots\ldots\ldots\ldots\ldots\ldots\ldots\ldots\ldots\ldots\ldots\ldots \\ K(x_m, y_1) & K(x_m, y_2) & \ldots & K(x_m, y_m) \end{vmatrix} \tag{4}$$

But, neglecting the divisor $m!$, the sum (3) multiplied by $(1/n)^m$ can be considered as an *approximate sum* for the (Riemann) m-fold integral

$$\int\int \ldots \int K \begin{pmatrix} \xi_1, & \xi_2, & \ldots, & \xi_m \\ \xi_1, & \xi_2, & \ldots, & \xi_m \end{pmatrix} d\xi_1 d\xi_2 \ldots d\xi_m$$

(all the integrations being between 0 and 1); hence the 'limit' of (2) as $n \to \infty$ is the elegant power series

$$\mathscr{D}(\lambda) = 1 + \sum_{m=1}^{\infty} \frac{(-\lambda)^m}{m!} \int\int \ldots \int K \begin{pmatrix} \xi_1, & \xi_2, & \ldots, & \xi_m \\ \xi_1, & \xi_2, & \ldots, & \xi_m \end{pmatrix} d\xi_1 d\xi_2 \ldots d\xi_m. \tag{5}$$

This is the basic function $\mathscr{D}(\lambda)$ of Fredholm.

Setting aside 'poetic license', we shall now show, with the help of Hadamard's theorem (see Appendix II), that the infinite series (5) is *everywhere convergent*, i.e. that $\mathscr{D}(\lambda)$ is an *entire function* of λ, provided that the given kernel $K(x, y)$ is bounded:

$$|K(x, y)| \leqslant N, \tag{6}$$

and, of course, integrable.

In fact, from Hadamard's theorem and (6) it follows that

$$\left| \int\int \ldots \int K \begin{pmatrix} \xi_1, & \xi_2, & \ldots, & \xi_m \\ \xi_1, & \xi_2, & \ldots, & \xi_m \end{pmatrix} d\xi_1 d\xi_2 \ldots d\xi_m \right| \leqslant \left| K \begin{pmatrix} \xi_1, & \ldots, & \xi_m \\ \xi_1, & \ldots, & \xi_m \end{pmatrix} \right| \leqslant N^m m^{\frac{1}{2}m};$$

consequently, series (5) has the majorant

$$\sum_{m=1}^{\infty} \frac{m^{\frac{1}{2}m}}{m!} (|\lambda| N)^m,$$

† Although it may seem superfluous at present to have *two* rows of variables in the symbol $K(::::)$, it will be useful later.

which is everywhere convergent because the power series

$$\Omega(z) = \sum_{m=1}^{\infty} a_m z^m \quad \text{with} \quad a_m = \frac{m^{\frac{1}{2}m}}{m!} \tag{6'}$$

has an *infinite* radius of convergence, for†

$$\lim_{m \to \infty} \frac{a_m}{a_{m+1}} = \lim_{m \to \infty} \frac{m^{\frac{1}{2}m}}{(m+1)^{\frac{1}{2}m}} \frac{m+1}{(m+1)^{\frac{1}{2}}}$$

$$= \lim_{m \to \infty} (m+1)^{\frac{1}{2}} \left[\left(1 + \frac{1}{m} \right)^m \right]^{-\frac{1}{2}} = \infty.$$

Now we return to the function $\mathscr{D}(x, y; \lambda)$ defined by (1) and observe that for $|\lambda| < N^{-1}$ we have

$$\mathscr{D}(x, y; \lambda) = \sum_{n=0}^{\infty} C_n(x, y) \frac{(-\lambda)^n}{n!}, \tag{7}$$

where $C_n(x, y)$ are certain suitable coefficients. In order to determine these coefficients, we shall use equations (2·1-10) for the resolvent kernel, which assumes now the form

$$\mathscr{D}(\lambda) K(x, y) + \mathscr{D}(x, y; \lambda) = \lambda \int K(x, z) \mathscr{D}(z, y; \lambda) \, dz \tag{8}$$

$$= \lambda \int \mathscr{D}(x, z; \lambda) K(z, y) \, dz,$$

because

$$H(x, y; \lambda) = \frac{\mathscr{D}(x, y; \lambda)}{\mathscr{D}(\lambda)}. \tag{9}$$

If we call I_m the coefficient of $(-\lambda)^m/m!$ in series (5), we obtain thus the recurrence relation

$$I_m K(x, y) + C_m(x, y) + m \int K(x, z) C_{m-1}(z, y) \, dz = 0 \quad (m = 1, 2, \ldots), \tag{10}$$

which, together with the obvious formula

$$C_0(x, y) = \mathscr{D}(0) H(x, y; 0) = -K(x, y), \tag{11}$$

determines potentially all the $C_m(x, y)$.

Let us now consider the m-fold integral

$$C_m^*(x, y) = \iint \ldots \int K \begin{pmatrix} x, \xi_1, \xi_2, \ldots, \xi_m \\ y, \xi_1, \xi_2, \ldots, \xi_m \end{pmatrix} d\xi_1 d\xi_2 \ldots d\xi_m.$$

† Remember that $\lim_{m \to \infty} \left(1 + \dfrac{1}{m} \right)^m = e$.

Developing the determinant by the elements of its first row, we obtain

$$C_m^* = \iint \dots \int \left\{ K(x,y)\, K\begin{pmatrix} \xi_1, \dots, \xi_m \\ \xi_1, \dots, \xi_m \end{pmatrix} \right.$$

$$\left. + \sum_{k=1}^{m} (-1)^k\, K(x,\xi_k)\, K\begin{pmatrix} \xi_1,\, \xi_2,\, \dotfill,\, \xi_m \\ y,\, \xi_1,\, \dots,\, \xi_{k-1},\, \xi_{k+1},\, \dots,\, \xi_m \end{pmatrix} \right\} d\xi_1 d\xi_2 \dots d\xi_m$$

$$= I_m K(x,y) - \sum_{k=1}^{m} \int K(x,\xi_k)\, d\xi_k$$

$$\times \iint \dots \int K\begin{pmatrix} \xi_k,\, \xi_1,\, \xi_2,\, \dots,\, \xi_{k-1},\, \xi_{k+1},\, \dots,\, \xi_m \\ y,\, \xi_1,\, \xi_2,\, \dots,\, \xi_{k-1},\, \xi_{k+1},\, \dots,\, \xi_m \end{pmatrix} d\xi_1 \dots d\xi_{k-1} d\xi_{k+1} \dots d\xi_m$$

$$= I_m K(x,y) - \sum_{k=1}^{m} \int K(x,\xi_k)\, C_{m-1}^*(\xi_k, y)\, d\xi_k;$$

but the last integral does not depend on k, since the integration variable can be called z instead of ξ_k without changing the value of the integral; hence we have

$$C_m^*(x,y) = I_m K(x,y) - m \int K(x,z)\, C_{m-1}^*(z,y)\, dz \quad (m = 1, 2, \dots).$$

Further, we also have

$$C_0^*(x,y) = K(x,y),$$

and, by comparing (10) and (11), we see that $C_m(x,y) = -C_m^*(x,y)$.

In other words, we find that *at least* for $|\lambda| < N^{-1}$ the function $\mathscr{D}(x,y;\lambda)$ has the elegant power expansion

$$\mathscr{D}(x,y;\lambda) = -\sum_{m=0}^{\infty} \frac{(-\lambda)^m}{m!} \iint \dots \int K\begin{pmatrix} x,\, \xi_1,\, \dots,\, \xi_m \\ y,\, \xi_1,\, \dots,\, \xi_m \end{pmatrix} d\xi_1 d\xi_2 \dots d\xi_m, \quad (12)$$

which is similar to (5).

Moreover, this expansion is not only valid for $|\lambda| < N^{-1}$ but for *any* λ, i.e. $\mathscr{D}(x,y;\lambda)$ is also an *entire* function of λ because the radius of convergence of the previous series is also *infinite*. Using Hadamard's theorem again, this can be immediately verified, since

$$\left| \iint \dots \int K\begin{pmatrix} x,\, \xi_1,\, \dots,\, \xi_m \\ y,\, \xi_1,\, \dots,\, \xi_m \end{pmatrix} d\xi_1 d\xi_2 \dots d\xi_m \right|$$

$$\leqslant \left| K\begin{pmatrix} x,\, \xi_1,\, \dots,\, \xi_m \\ y,\, \xi_1,\, \dots,\, \xi_m \end{pmatrix} \right| \leqslant N^{m+1}(m+1)^{\frac{1}{2}(m+1)}.$$

Consequently, series (12) has the majorant

$$N \sum_{m=0}^{\infty} a'_m(|\lambda|N)^m \quad \text{with} \quad a'_m = \frac{1}{m!}(m+1)^{\frac{1}{2}(m+1)},$$

but

$$\rho_m = \frac{a'_m}{a'_{m+1}} = (m+1)^{\frac{1}{2}} \left(\frac{m+1}{m+2}\right)^{\frac{1}{2}} \left[\left(1+\frac{1}{m+1}\right)^{m+1}\right]^{-\frac{1}{2}};$$

hence the limit of ρ_m as $m \to \infty$ is *infinite*, etc.

The *Fredholm formulae* (5) and (12) give us the resolvent kernel $H(x, y; \lambda)$ in the whole λ-plane, in the elegant form (9) which shows that this resolvent kernel is a *meromorphic* function of λ.† More-over, according to (9), *for any λ which is not an eigenvalue*, the unique solution of a Fredholm integral equation of the second kind can be put into the form

$$\phi(x) \equiv f(x) - \frac{\lambda}{\mathscr{D}(\lambda)} \int \mathscr{D}(x, y; \lambda) f(y) \, dy. \tag{13}$$

The sole inconvenience is that the infinite series (5) and (12), in spite of their good convergence, are too complicated for numerical calculations, due to the multiple integrals appearing in the various terms. However, other methods do exist for numerical calculations, and we shall soon say something about them (§ 2·6).

Among other things, a simple, interesting relation between the derivatives with respect to λ of $\mathscr{D}(\lambda)$ and $\mathscr{D}(x, y; \lambda)$ can be deduced from the Fredholm formulae. To be precise, since differentiation term-by-term is permitted, we obtain

$$\mathscr{D}^{(n)}(x, y; \lambda) \equiv \frac{\partial^n}{\partial \lambda^n} \mathscr{D}(x, y; \lambda)$$

$$= (-1)^{n+1} \sum_{m=n}^{\infty} \frac{(-\lambda)^{m-n}}{(m-n)!} \int\!\!\int \cdots \int K \begin{pmatrix} x, \xi_1, \ldots, \xi_m \\ y, \xi_1, \ldots, \xi_m \end{pmatrix} d\xi_1 d\xi_2 \ldots d\xi_m;$$

consequently, for $x = y$, we have

$$\int \mathscr{D}^{(n)}(x, x; \lambda) \, dx = (-1)^{n+1} \sum_{m=n}^{\infty} \frac{(-\lambda)^{m-n}}{(m-n)!}$$

$$\times \int\!\!\int \cdots \int K \begin{pmatrix} x, \xi_1, \ldots, \xi_m \\ x, \xi_1, \ldots, \xi_m \end{pmatrix} dx \, d\xi_1 \ldots d\xi_m. \tag{14}$$

---

† A *meromorphic* function is the quotient of two analytic *entire* functions. In any finite portion of the complex plane it possesses only poles; hence, it possesses only a finite number of such poles in any finite portion of the λ-plane.

On the other hand, (5) can be written in the form

$$\mathscr{D}(\lambda) = 1 + \sum_{m=0}^{\infty} \frac{(-\lambda)^{m+1}}{(m+1)!} \int\int\ldots\int K\begin{pmatrix}\xi_0, \xi_1, \ldots, \xi_m \\ \xi_0, \xi_1, \ldots, \xi_m\end{pmatrix} d\xi_0 d\xi_1 \ldots d\xi_m,$$

and with $(n+1)$ differentiations with respect to λ, we obtain

$$\mathscr{D}^{(n+1)}(\lambda) = (-1)^{n+1} \sum_{m=n}^{\infty} \frac{(-\lambda)^{m-n}}{(m-n)!}$$

$$\times \int\int\ldots\int K\begin{pmatrix}\xi_0, \xi_1, \xi_2, \ldots, \xi_m \\ \xi_0, \xi_1, \xi_2, \ldots, \xi_m\end{pmatrix} d\xi_0 d\xi_1 \ldots d\xi_m. \quad (15)$$

Now the right-hand sides of (14) and (15) differ only in that one has the variable of integration x, while the other has ξ_0; hence

$$\mathscr{D}^{(n+1)}(\lambda) = \int \mathscr{D}^{(n)}(x, x; \lambda)\, dx \quad (n = 0, 1, 2, \ldots). \quad (16)$$

In particular, for $n = 0$ and $|\lambda| < N^{-1}$, we have

$$\frac{\mathscr{D}'(\lambda)}{\mathscr{D}(\lambda)} = \int H(x, x; \lambda)\, dx = - \sum_{n=1}^{\infty} \lambda^{n-1} \int K_n(x, x)\, dx.$$

Now if we put†

$$\int K_n(x, x)\, dx = A_n \quad (n = 1, 2, 3, \ldots), \quad (17)$$

we see that the *logarithmic derivative* of $\mathscr{D}(\lambda)$ has the simple and important power expansion

$$\frac{\mathscr{D}'(\lambda)}{\mathscr{D}(\lambda)} = - \sum_{n=0}^{\infty} A_{n+1} \lambda^n, \quad (18)$$

whose radius of convergence obviously coincides with the minimal distance of the eigenvalues from the origin.

Among other things this expansion can be utilized to show that *a Volterra integral equation has no eigenvalues*, as has been seen implicitly in § 1·5.

In fact, if $\qquad K(x, y) \equiv 0 \quad$ for $\quad y > x,$

then $\qquad K_n(x, y) \equiv 0 \quad$ for $\quad y > x,\ n = 2, 3, \ldots,$

and consequently

$$K(x, z)\, K_n(z, x) \equiv 0 \quad (x \neq z,\ n = 1, 2, 3, \ldots).$$

† The constants A_n are often called the *traces* of the kernel $K(x, y)$. (German: die *Spuren*.)

But this shows that $A_2 = A_3 = \ldots = 0$; hence

$$\frac{\mathscr{D}'(\lambda)}{\mathscr{D}(\lambda)} = -A_1;$$

and, considering that $\mathscr{D}(0) = 1$, we have

$$\mathscr{D}(\lambda) = e^{-A_1 \lambda}. \tag{18'}$$

This shows that in this case there are no eigenvalues.

Another important consequence of the earlier Fredholm formulae is the possibility of determining explicitly the eigenfunctions corresponding to an eigenvalue λ_0 of index $r = 1$ for which the function $\mathscr{D}(x, y; \lambda_0)$ does not vanish identically.† For, the first equation (8), which in the case $\lambda = \lambda_0$ assumes the form

$$\mathscr{D}(x, y; \lambda_0) - \lambda_0 \int K(x, z)\, \mathscr{D}(z, y; \lambda_0)\, dz = 0,$$

shows that an eigenfunction corresponding to λ_0 is

$$\phi_0(x) = C\mathscr{D}(x, y_0; \lambda_0), \tag{19}$$

where C and y_0 are two arbitrary constants. But we already know that all the eigenfunctions must be contained in a family of type (2·3-15) with actually $r = 1$; hence a change in y_0 can only multiply $\mathscr{D}(x, y_0; \lambda_0)$ by a factor independent of x. In other words, we have necessarily

$$\frac{\mathscr{D}(x, y_0; \lambda_0)}{\mathscr{D}(x, y_1; \lambda_0)} = \text{const.}, \tag{20}$$

and this shows that $\mathscr{D}(x, y; \lambda_0)$ is necessarily of the form

$$\mathscr{D}(x, y; \lambda_0) = F(x)\, G(y). \tag{21}$$

If the eigenvalue λ_0 has the rank $r > 1$, then $\mathscr{D}(x, y; \lambda)$ vanishes identically and, for $k < r$, this is true also for the more general entire functions

$$-\mathscr{D}(x_1, \ldots, x_k;\ y_1, \ldots, y_k;\ \lambda)$$
$$= \sum_{m=0}^{\infty} \frac{(-\lambda)^m}{m!} \int\!\int \ldots \int K \begin{pmatrix} x_1, \ldots, x_k, \xi_1, \ldots, \xi_m \\ y_1, \ldots, y_k, \xi_1, \ldots, \xi_m \end{pmatrix} d\xi_1 d\xi_2 \ldots d\xi_m, \tag{22}$$

† If the multiplicity m of the eigenvalue λ_0 is also equal to *one*, i.e. if $\mathscr{D}'(\lambda_0) \neq 0$, this is an immediate consequence of (16) for $n = 1$ because the identical vanishing of $\mathscr{D}(x, y; \lambda_0)$ implies $\mathscr{D}'(\lambda_0) = 0$; but if $r = 1$, $m > 1$ the proof requires the consideration of the functions $\mathscr{D}(x_1, x_2, \ldots;\ y_1, y_2, \ldots;\ \lambda)$ given by (22).

which generalize the previous function $\mathscr{D}(x, y; \lambda)$. But in such a case the function $\mathscr{D}(x_1, ..., x_r; y_1, ..., y_r; \lambda)$ corresponding to $k = r$ does not vanish identically, and it furnishes the r linearly independent eigenfunctions

$$\phi_{0h}(x) \equiv \mathscr{D}(x_1^{(0)}, ..., x_{h-1}^{(0)}, x, x_{h+1}^{(0)}, ..., x_r^{(0)}; y_1^{(0)}, ..., y_r^{(0)}; \lambda_0)$$
$$(h = 1, 2, ..., r). \quad (23)$$

For more particulars, see, for example, Lovitt [30].

Finally, we observe that as in the case of the Volterra integral equation (§ 1·12), the Fredholm equation of the second kind can be immediately reduced to

$$\phi(x) - \lambda^n \int K_n(x, y) \, \phi(y) \, dy = f_n(x) \quad (n = 1, 2, 3, ...), \quad (24)$$

with the iterated kernel $K_n(x, y)$, where

$$f_1(x) \equiv f(x), \quad f_{n+1}(x) = f(x) + \lambda \int K(x, y) f_n(y) \, dy \quad (n = 1, 2, 3, ...).$$

This transformation also shows that *if λ_0 is the eigenvalue corresponding to the eigenfunction $\phi_0(x)$ of the kernel $K(x, y)$, then $\phi_0(x)$ is also an eigenfunction of the iterated kernel $K_n(x, y)$, and λ_0^n is the corresponding eigenvalue.*

In order to invert this theorem, we first observe that if $H_n(x, y; \lambda)$ is the resolvent kernel corresponding to the iterated kernel $K_n(x, y)$, then the two power expansions

$$- \lambda H(x, y; \lambda) = \lambda K(x, y) + \lambda^2 K_2(x, y) + ...,$$
$$- \lambda^n H_n(x, y; \lambda^n) = \lambda^n K_n(x, y) + \lambda^{2n} K_{2n}(x, y) + ...,$$

are analogous to the expansions of the two functions $f(z)$ and $f_{m,1}(z)$ of § 1·9.† Consequently, in view of (1·9-18), there are certainly n constants $\eta_1, \eta_2, ..., \eta_n$ (depending only on the n complex roots of unity $\epsilon_1, \epsilon_2, ..., \epsilon_n$) such that

$$\lambda^{n-1} H_n(x, y; \lambda^n)$$
$$= \eta_1 H(x, y; \epsilon_1 \lambda) + \eta_2 H(x, y; \epsilon_2 \lambda) + ... + \eta_n H(x, y; \epsilon_n \lambda) \quad (25)$$

identically.

But the identity (25) shows that the point $\mu = \mu_0$ is a singular point of $H_n(x, y; \mu)$ if *and only if* at least one of the n points

$$\epsilon_1 \sqrt[n]{\mu_0}, \quad \epsilon_2 \sqrt[n]{\mu_0}, \quad ..., \quad \epsilon_n \sqrt[n]{\mu_0},$$

† Change m into n and put $a_0 = 0$.

where $\sqrt[n]{\mu_0}$ denotes any fixed nth root of μ_0, is a singular point of $H(x, y; \lambda)$. Hence:

If μ_0 is an eigenvalue of the iterated kernel $K_n(x, y)$, at least one of the n complex nth roots of μ_0 is an eigenvalue of the original kernel $K(x, y)$.

In particular, in the case $n = 2$ we obtain from (25)

$$H_2(x, y; \lambda^2) = \frac{1}{2\lambda} [H(x, y; \lambda) - H(x, y; -\lambda)]. \qquad (26)$$

2·6. Numerical Solution of Integral Equations

As already noted, the Fredholm formulae are of little use in the *numerical solution* of integral equations. However, in most cases the kernel can be approximated by a suitable PG-kernel or a step-function; then, as we have already seen, the integral equation can be reduced to an algebraic system of linear equations.

When a PG-kernel is used, the most important step is to choose the $X_k(x)$ and $Y_k(y)$ so judiciously that a good approximation is given by only a few terms. Sometimes two or three terms are sufficient; unfortunately, there are few general rules to facilitate this choice. For instance, if $K(x, y)$ vanishes identically for some special value of x or y, it may be useful for the functions X_k or Y_k to be zero at such points. In any case, when we study *orthogonal systems* in the next chapter (see § 3·6) we shall find a general method of producing an infinite number of PG-kernels which approximate a given L_2-kernel '*in the mean*'.

On the contrary, if use is made of step-functions (i.e. if the integral of the given integral equation is replaced by a suitable finite sum), then instead of dividing the basic interval into subintervals of the same size as in § 1·2, it may be useful to divide it according to the zeros of a certain polynomial of Legendre. This method was used by Gauss for numerical integration, and was developed by Nyström.†

In both cases, if the given kernel $K(x, y)$ is approximated by another kernel $K^*(x, y)$ with a known maximal (pointwise) error:

$$| K(x, y) - K^*(x, y) | < \epsilon, \qquad (1)$$

† 'Über die praktische Auflösung von linearen Integralgleichungen', *Comm. Phys. Math. Soc. Sci. Fenn.* IV, 15 and V, 5; *Acta Math.* **56**, 1931.

then it may be useful to have an upper bound for the maximal error $|\phi(x) - \phi^*(x)|$ of the corresponding solutions of the Fredholm integral equation of the second kind. Many years ago I gave† an explicit formula for this error

$$|\phi(x) - \phi^*(x)|$$

$$< \epsilon \,|\, \lambda \,|\, \frac{\Omega(L)\,\Omega'(L) + L[\Omega'^2(L) + \Omega(L)\,\Omega''(L)]}{|\,\mathscr{D}^*(\lambda)\,|\,[\,|\,\mathscr{D}^*(\lambda)\,| - \epsilon\,|\,\lambda\,|\,\Omega'(L)]}\,\max |f(x)|, \quad (2)$$

where Ω is the function of (2·5-6′), $\mathscr{D}^*(\lambda)$ is the Fredholm function (2·5-5) for the approximate kernel $K^*(x, y)$, and

$$L = |\,\lambda\,|\,[\max|\,K(x, y)\,| + \epsilon]. \quad (3)$$

It is even possible to replace the function $\Omega(z)$ and its derivatives by the corresponding derivatives of the elementary function

$$\Omega_0(z) = (1 + z)\exp(\tfrac{1}{2}ez^2), \quad (4)$$

which is a *majorant* (in Cauchy's sense) of the function $\Omega(z)$.

Another method for the numerical solution of the Fredholm integral equation is furnished by Enskog's method, which we shall study in the following chapter (§ 3·7) because it is closely connected with the theory of orthogonal functions.

Furthermore, C. Lancloz‡ has recently developed a method of 'minimized iteration', which in some instances seems to give very good numerical results. This method can be interpreted as the application of a suitable *summation process* to the Neumann series (2·1-5). Here one replaces the Neumann series by another series which has the same sum as the first, when the first is convergent; however it remains convergent even in some cases in which the first series diverges.§

2·7. The Fredholm Solution of the Dirichlet Problem

The most important '*theoretical*' application of the theory of integral equations is the derivation of *existence theorems* from the

† F. Tricomi, 'Sulla risoluzione numerica delle equazioni integrali di Fredholm', *Rend. Accad. Lincei* (5), **33** (I), pp. 483–6 and **33** (II), pp. 26–30, 1924.

‡ C. Lancloz, 'An iteration method for the solution of the eigenvalue problem of linear differential and integral operators', *J. Research, National Bur. of Standards*, **45**, 1950, pp. 255–282.

§ See also the recent book of Bückner [3].

corresponding *uniqueness theorems*; the latter are generally easier to prove.

In fact, if a problem concerning a certain function ϕ can be reduced to a Fredholm integral equation of the second kind with an L_2-kernel, then the *alternative theorem* of § 2·3 assures us of the *existence* of such a function, provided that λ is not an eigenvalue, i.e. provided that the corresponding homogeneous equation has only the trivial solution $\phi \equiv 0$, in other words, provided that a *uniqueness theorem* is valid for our problem.

This general principle is the key to the celebrated Fredholm solution (1900) of the Dirichlet problem; this solution first attracted the mathematical world to the theory of integral equations.

In its simplest form, the Dirichlet problem is the determination of a *harmonic function* $u(x, y)$ of two variables, within a given two-dimensional domain D; the values u_s of the function are prescribed on the boundary s of D. A harmonic function is a solution of Laplace's equation

$$\frac{\partial^2 u}{\partial x^2} + \frac{\partial^2 u}{\partial y^2} = 0, \tag{1}$$

which is 'regular' in the closed domain D.

Here for the sake of simplicity 'regular' will mean 'continuous together with its first and second derivatives'.

The *uniqueness* of the solution of the problem, i.e. the fact that $u \equiv 0$ in the whole domain D if $u_s \equiv 0$ on the boundary, is almost obvious. In fact, this is an immediate consequence of the elementary identity

$$\iint_D \left[\left(\frac{\partial u}{\partial x}\right)^2 + \left(\frac{\partial u}{\partial y}\right)^2 \right] dx\,dy = -\oint_s u_s \frac{du}{dn} ds, \tag{2}$$

where du/dn denotes the derivative of u along the *inner normal* of s.†

However, the proof of the *existence* of the harmonic function u is very difficult, and before Fredholm it had been rigorously achieved only for a relatively restricted class of domains D.

As we have already explained, the key to the Fredholm method is the reduction of the determination of u to the solution of a Fredholm integral equation of the second kind. Then we do not really need to solve this equation, for the alternative and uniqueness theorems immediately give us the existence of the harmonic

† Or, even better, we can use the fact that the extrema of a harmonic function must lie on the boundary.

function $u(x, y)$ under very general hypotheses about the boundary s of D and the given values u_s of u there.

Without going into details,† we shall discuss here only the main device for the reduction of the Dirichlet problem to an integral equation, which is to consider the unknown function u as a *logarithmic potential of a double layer*, due to a distribution with a suitable (unknown) density $\mu(s)$ on the boundary s of D:

$$u(x, y) = -\int_0^l \mu(s)\frac{d\log r}{dn}\, ds. \tag{3}$$

Here s is the *arc length* along the boundary (from an arbitrary origin), l the total length of this boundary, and

$$r = r(P, Q) = \sqrt{[(x - x_s)^2 + (y - y_s)^2]}$$

is the distance between the generic point $P \equiv (x, y)$ of D and the generic point $Q \equiv (x_s, y_s)$ of the boundary.

A well-known property of the potential of a double layer is that if P approaches any point Q_0 $(s = s_0)$ of the boundary *from the interior of D* we have

$$\lim_{P \to Q_0} u = -\int_0^l \mu(s)\frac{d}{dn}\log r(Q_0, Q)\, ds + \pi\mu(s_0). \tag{4}$$

Consequently, if we call $f(s)$ the given values u_s of u on the boundary, then we must have on this curve

$$\pi\mu(s_0) - \int_0^l \frac{d}{dn}\log r(Q_0, Q)\mu(s)\, ds = f(s) \quad (0 \leqslant s \leqslant l),$$

which is a Fredholm integral equation of the second kind in the unknown $\mu(s)$. Even better, if we suppose $l = 1$, and put

$$\frac{d}{dn}\log r(Q_0, Q) = K(s_0, s), \tag{5}$$

writing x and y instead of s_0 and s respectively, then we have the integral equation

$$\mu(x) - \frac{1}{\pi}\int_0^1 K(x, y)\mu(y)\, dy = \frac{1}{\pi}f(x). \tag{6}$$

Realizing that a potential of double (as well as one of single) layer is identically zero inside of D if and only if the corresponding

---

† See, for example, Lovitt [30] or O. D. Kellogg, *Foundations of Potential Theory*, Springer, Berlin, 1929, Ch. XI.

density μ vanishes identically, we can assert now without more ado that *the Dirichlet problem is solvable for at least all the domains D whose boundaries give rise to a continuous, or at least a quadratically integrable, kernel* (5).

Supposing that the boundary of D is represented by the parametric equations

$$x = \alpha(s), \quad y = \beta(s) \quad (0 \leqslant s \leqslant 1), \tag{7}$$

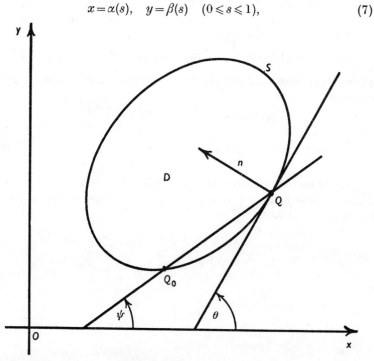

Fig. 5

where s again denotes the arc length, we readily obtain the formula

$$-K(x,y) = \frac{\sin(\theta - \psi)}{r(Q_0, Q)} = \frac{\partial}{\partial y} \operatorname{arctg} \frac{\beta(y) - \beta(x)}{\alpha(y) - \alpha(x)}$$

$$= \frac{[\alpha(y) - \alpha(x)]\beta'(y) - [\beta(y) - \beta(x)]\alpha'(y)}{[\alpha(y) - \alpha(x)]^2 + [\beta(y) - \beta(x)]^2},$$

where θ and ψ denote respectively the angles which the tangent on the curve at Q and the secant $Q_0 Q$ form with the x-axis. But if the functions α and β are differentiable at least twice, we can write

$$\alpha(x) = \alpha(y) + (x-y)\,\alpha'(y) + \tfrac{1}{2}(x-y)^2\,[\alpha''(y) + \epsilon_1],$$
$$\beta(x) = \beta(y) + (x-y)\,\beta'(y) + \tfrac{1}{2}(x-y)^2\,[\beta''(y) + \epsilon_2],$$

where $\lim \epsilon_1 = \lim \epsilon_2 = 0$ as $x \to y$; hence

$$[\alpha(y) - \alpha(x)]\,\beta'(y) - [\beta(y) - \beta(x)]\,\alpha'(y)$$
$$= \tfrac{1}{2}[\gamma(y) + \epsilon_2\alpha'(y) - \epsilon_1\beta'(y)]\,(x-y)^2,$$
$$[\alpha(y) - \alpha(x)]^2 + [\beta(y) - \beta(x)]^2$$
$$= \{1 + (x-y)\,[\epsilon_1\alpha'(y) + \epsilon_2\beta'(y)] + O[(x-y)^2]\}\,(x-y)^2,$$

where $\gamma(y)$ denotes the *curvature* of our curve s at Q. Consequently we have

$$-K(x,y) = \frac{1}{2}\frac{\gamma(y) + \epsilon_2\alpha'(y) - \epsilon_1\beta'(y)}{1 + O(x-y),} \,, \tag{8}$$

and, in particular, $\quad \lim\limits_{x \to y} K(x,y) = -\tfrac{1}{2}\gamma(y).$ $\qquad\qquad$ (9)

This shows that the *Dirichlet problem can be solved at least for any domain whose boundary is a curve having a tangent and a finite curvature everywhere.*

CHAPTER III

Symmetric Kernels and Orthogonal Systems of Functions

3·1. Introductory Remarks and a Process of Orthogonalization

For a *symmetric kernel*, i.e. for a kernel for which†

$$K(x, y) = K(y, x), \qquad (1)$$

the associated eigenfunctions ψ_h coincide with the proper eigenfunctions ϕ_h. It follows from the orthogonality property (§ 2·3) that any pair $\phi_h(x)$, $\phi_k(x)$ of eigenfunctions of a symmetric kernel, corresponding to two *different* eigenvalues λ_h, λ_k, satisfy a similar orthogonality condition

$$(\phi_h, \phi_k) \equiv \int_a^b \phi_h(x) \, \phi_k(x) \, dx = 0 \qquad (h \neq k) \qquad (2)$$

in the basic interval (a, b).‡

Because of this connection between symmetric kernels and *orthogonal systems of functions*, i.e. (finite or infinite) systems of L_2-functions

$$\{\phi_h\} \equiv \phi_1(x), \phi_2(x), \phi_3(x), \dots \qquad (3)$$

which satisfy (2),§ it is suitable to begin our study of the symmetric

---

† We shall consider only *real functions*, even though most of our remarks can be readily extended to complex functions. Note, however, that the analogue of the symmetric kernel for complex functions is the *Hermitian kernel*: $K(y, x) = \bar{K}(x, y)$, where the bar denotes the conjugate of a complex number ($\bar{z} = x - iy$ if $z = x + iy$).

‡ In this chapter we shall use (a, b) as our basic interval instead of $(0, 1)$ because most of our proofs will remain valid even if the interval is *infinite* (i.e. even if $b = \infty$, or $a = -\infty$, or $a = -\infty$ and $b = \infty$) provided all the integrals converge.

§ We can denote the functions of an orthogonal system by $\phi_1, \phi_2, \phi_3, \dots$ because under very general conditions, any orthogonal system is either *finite* (i.e. contains a finite number of functions) or it is *denumerable*. See, for example, Tricomi [46], p. 23.

Fredholm integral equation with a brief survey (§§ 3·1–3·5) of the theory of orthogonal functions.†

We shall assume throughout that every function ϕ_h of our othogonal systems is an L_2-function which does not vanish almost everywhere, i.e. that

$$\| \phi_h \|^2 = \int_a^b \phi_h^2(x)\, dx > 0. \tag{4}$$

Thus we can suppose that our functions are not only orthogonalized but also *normalized*, i.e. that

$$(\phi_h, \phi_k) = \begin{cases} 0 & (h \neq k), \\ 1 & (h = k). \end{cases} \tag{5}$$

Such a system will be called *orthonormal* (or briefly, an ON-system).

The functions of any orthogonal system (briefly, an O-system) are *linearly independent*; for, if there exist constants $c_1, c_2, \ldots, c_n$ which are not all zero and are such that

$$c_1 \phi_1(x) + c_2 \phi_2(x) + \ldots + c_n \phi_n(x) \equiv 0 \tag{6}$$

almost everywhere in the basic interval (a, b), then multiplying by $\phi_h(x)$ $(h = 1, 2, \ldots, n)$ and integrating over (a, b), we have

$$c_h \int_a^b \phi_h^2(x)\, dx = 0,$$

which, by (4), implies that $c_h = 0$, i.e. $c_1 = c_2 = \ldots = c_n = 0$.

It is amazing that linear independence is not only *necessary* for orthogonality, but, in a certain sense, is also *sufficient*. This is clear because we can always use the following procedure:

PROCESS OF ORTHOGONALIZATION. *Given any (finite or denumerable) system of linearly independent L_2-functions*

$$\psi_1(x), \quad \psi_2(x), \quad \psi_3(x), \quad \ldots,$$

---

† This theory, which includes Fourier series and orthogonal polynomials, is one of the most important tools of modern analysis. A good book about the *general* theory is Kaczmarz-Steinhaus [23]. My book [47] deals briefly (56 pp.) with the general theory and more extensively with trigonometric series and orthogonal polynomials.

it is always possible to find constants h_{rs} such that the functions

$$\left.\begin{aligned}
&\phi_1(x) = \psi_1(x),\\
&\phi_2(x) = h_{21}\psi_1(x) + \psi_2(x),\\
&\phi_3(x) = h_{31}\psi_1(x) + h_{32}\psi_2(x) + \psi_3(x),\\
&\dotfill,\\
&\phi_n(x) = h_{n1}\psi_1(x) + h_{n2}\psi_2(x) + \dots + h_{nn-1}\psi_{n-1}(x) + \psi_n(x),\\
&\dotfill,
\end{aligned}\right\} \quad (7)$$

are orthogonal in the basic interval (a, b).

We prove this by mathematical induction. Observe first that the system (7) can be readily resolved with respect to the functions $\psi_1, \psi_2, \dots$, i.e. it can be put into the equivalent form

$$\begin{aligned}
&\phi_1(x) = \psi_1(x),\\
&\phi_2(x) = k_{21}\phi_1(x) + \psi_2(x),\\
&\phi_3(x) = k_{31}\phi_1(x) + k_{32}\phi_2(x) + \psi_3(x),\\
&\dotfill,\\
&\phi_n(x) = k_{n1}\phi_1(x) + k_{n2}\phi_2(x) + \dots + k_{nn-1}\phi_{n-1}(x) + \psi_n(x),\\
&\dotfill
\end{aligned}$$

We shall suppose that for $n-1$ functions the coefficients k_{rs} have already been determined; i.e. we know k_{rs} for

$$1 \leqslant s < r \leqslant n-1.$$

We shall now show that the coefficients for the nth function k_{ns} $(s = 1, 2, \dots, n-1)$ can be readily calculated. In fact, from the $n-1$ conditions

$$\begin{aligned}
0 = (\phi_n, \phi_s) &= k_{n1}(\phi_1, \phi_s) + k_{n2}(\phi_2, \phi_s) + \dots + k_{nn-1}(\phi_{n-1}, \phi_s) + (\psi_n, \phi_s)\\
&= k_{ns}(\phi_s, \phi_s) + (\psi_n, \phi_s) \quad (s = 1, 2, \dots, n-1),
\end{aligned}$$

we get
$$k_{ns} = -\frac{(\psi_n, \phi_s)}{(\phi_s, \phi_s)}. \quad (8)$$

These coefficients are well defined; $(\phi_s, \phi_s) \neq 0$ because ϕ_s is a linear combination of the *linearly independent* functions $\psi_1, \psi_2, \dots, \psi_n$ and hence cannot be equal to zero almost everywhere.

3·2. Approximation and Convergence in the Mean

Let us consider a certain ON-system of functions $\{\phi_h\}$ and an 'arbitrary' function $f(x)$ defined on the basic interval of the system. Can the function $f(x)$ be 'represented' by a suitable linear combination

$$c_1 \phi_1(x) + c_2 \phi_2(x) + \ldots \tag{1}$$

of the functions of the ON-system? Of course the answer depends on the meaning which we give to the word *represented* and to the word *arbitrary* (function). Neglecting the less interesting case of finite systems, we could require that the infinite series $\Sigma c_h \phi_h(x)$ converge in the usual sense and that its sum coincide with $f(x)$. This leads to the fundamental, but difficult, theory of Fourier series and to other branches of modern analysis (series of orthogonal polynomials, etc.). However, less difficult results of another important type can be obtained if for the condition

$$\lim_{n \to \infty} \left[f(x) - \sum_{h=1}^{n} c_h \phi_h(x) \right] = 0 \tag{2}$$

we substitute the weaker† one

$$\lim_{n \to \infty} \int_a^b \left[f(x) - \sum_{h=1}^{n} c_h \phi_h(x) \right]^2 dx = 0. \tag{3}$$

Condition (3) states that to any positive ϵ there corresponds an index n_0 such that

$$\int_a^b \left[f(x) - \sum_{h=1}^{n} c_h \phi_h(x) \right]^2 dx < \epsilon \quad (n > n_0). \tag{4}$$

In such a case we shall say that the infinite series *converges in the mean* to the function $f(x)$ and we shall write

$$f(x) = \text{l.i.m.} \sum_{n \to \infty}^{n}_{h=1} c_h \phi_h(x). \tag{5}$$

It is of basic importance that convergence in the mean can be asserted under very general conditions; one condition is obviously that $f(x)$ be quadratically integrable, another is that the system $\{\phi_h\}$ be complete (a term we shall explain soon).

Using the orthonormality conditions (3·1-5), we have

$$I_n \equiv \int_a^b \left[f(x) - \sum_{h=1}^{n} c_h \phi_h(x) \right]^2 dx$$
$$= \int_a^b f^2(x)\, dx - 2 \sum_{h=1}^{n} c_h \int_a^b f(x)\, \phi_h(x)\, dx + \sum_{h=1}^{n} c_h^2.$$

† The justification for this adjective is the fact that if (2) is valid *uniformly* in the finite interval (a, b), then (3) follows, but not conversely. However, it may be that (2) is valid *not uniformly* and (3) is *not valid* at all.

If we denote by a_n the *Fourier coefficients* of the function $f(x)$ with respect to the system $\{\phi_h\}$, i.e. if

$$a_h = \int_a^b f(x)\,\phi_h(x)\,dx \quad (h = 1, 2, 3, \ldots), \tag{6}$$

we have
$$I_n = \int_a^b f^2(x)\,dx - 2\sum_{h=1}^n a_h c_h + \sum_{h=1}^n c_h^2$$

$$= \int_a^b f^2(x)\,dx - \sum_{h=1}^n a_h^2 + \sum_{h=1}^n (a_h - c_h)^2.$$

Since the quantity $\quad \int_a^b f^2(x)\,dx - \sum\limits_{h=1}^n a_h^2$

is independent of the choice of the coefficients c_h of the linear combination (1), and since the sum involving the square of the difference $a_h - c_h$ is never negative, we see that *for a given system* $\{\phi_h\}$, *a given function* $f(x)$, *and a given number n of terms of the linear combination* (1), *the non-negative integral* I_n *attains its minimal value if and only if the coefficients* c_h *coincide with the corresponding Fourier coefficients* (6) *of the function* $f(x)$.

Moreover, this minimal value of I_n, which we shall denote by I_n^*, is given by the formula

$$I_n^* = \int_a^b f^2(x)\,dx - \sum_{h=1}^n a_h^2. \tag{7}$$

Thus, for *any* L_2-function $f(x)$ and *any* positive integer n, *Bessel's inequality*

$$\sum_{h=1}^n a_h^2 \leqslant \int_a^b f^2(x)\,dx \tag{8}$$

holds. Hence the infinite series with non-negative terms Σa_h^2 is *always convergent*, and formula (7) can be put into the form

$$I_n^* = \left[\int_a^b f^2(x)\,dx - \sum_{h=1}^\infty a_h^2\right] + \sum_{h=n+1}^\infty a_h^2. \tag{9}$$

Thus we see that *a necessary and sufficient condition for an* L_2-*function* $f(x)$ *to be approximated in the mean by a linear combination of functions of a given ON-system is that Parseval's equation*

$$\sum_{h=1}^\infty a_h^2 = \int_a^b f^2(x)\,dx \tag{10}$$

hold for the given $f(x)$.

In fact, if
$$\Delta = \int_a^b f^2(x)\, dx - \sum_{h=1}^{\infty} a_h^2 \tag{11}$$

has a *positive* value η, we have $I_n^* \geqslant \eta$ for every n, and a *fortiori* $I_n \geqslant \eta$. However, if $\Delta = 0$ and $c_h = a_h$, the integral I_n^* coincides with

$$\sum_{h=n+1}^{\infty} a_h^2,$$

and hence can be made less than any positive ϵ provided that n is larger than a suitable n_0.

It is now easy to see the significance of the following:

DEFINITION. *An ON-system of functions is called complete in the whole class L_2 or in a subset M of this class, if Parseval's equation (10) holds for any function $f(x)$ of L_2 or of M, respectively.*

If a certain ON-system is complete in M, *any* function of M can be *approximated* in the mean arbitrarily closely by linear combinations of functions of the system and vice versa.

Suppose a function $f(x)$ is given as the *limit in the mean* of a certain sequence of functions $f_1(x), f_2(x), \ldots$; then it is (obviously) not determined at *every point*, but it is *unique* in the sense that if

$$f(x) = \underset{n \to \infty}{\text{l.i.m.}}\, f_n(x), \quad f^*(x) = \underset{n \to \infty}{\text{l.i.m.}}\, f_n(x), \tag{12}$$

then *the difference $f(x) - f^*(x)$ vanishes almost everywhere* in the basic interval (a, b).

In fact,
$$[f(x) - f^*(x)]^2 = \{[f(x) - f_n(x)] - [f^*(x) - f_n(x)]\}^2$$
$$\leqslant 2\{[f(x) - f_n(x)]^2 + [f^*(x) - f_n(x)]^2\},$$

and consequently
$$\int_a^b [f(x) - f^*(x)]^2\, dx \leqslant 2 \int_a^b [f(x) - f_n(x)]^2\, dx + 2 \int_a^b [f^*(x) - f_n(x)]^2\, dx.$$

Now from hypothesis (12) it follows that to any $\epsilon > 0$ there corresponds a positive integer n_0 such that for $n > n_0$ both integrals on the right-hand side are less than ϵ; hence if $n > n_0$

$$\int_a^b [f(x) - f^*(x)]^2\, dx < 4\epsilon.$$

Since the integral on the left-hand side is independent of n, it must be that
$$\int_a^b [f(x) - f^*(x)]^2\, dx = 0.$$

This shows that the difference $f(x) - f^*(x)$ vanishes almost everywhere in (a, b).

Finally, we shall show that Parseval's equation (10) can be generalized to obtain a formula which gives the integral of the *product* of two L_2-functions $f(x)$ and $g(x)$.

With the help of the above inequality for the square of a sum, it is easy to see that *as long as Parseval's equation holds for both functions $f(x)$ and $g(x)$, it also holds for any linear combination* $\lambda f(x) + \mu g(x)$. Comparing successively the coefficients of $2\lambda\mu$ on both sides of the equality

$$\int_a^b [\lambda f(x) + \mu g(x)]^2 \, dx = \sum_{h=1}^{\infty} (\lambda a_h + \mu b_h)^2,$$

where a_h and b_h are the Fourier coefficients of $f(x)$ and $g(x)$, respectively, we obtain the *generalized Parseval equation*

$$\int_a^b f(x) g(x) \, dx = \sum_{h=1}^{\infty} a_h b_h. \tag{13}$$

This equality, which can also be written in the form

$$\int_a^b f(x) g(x) \, dx = \sum_{h=1}^{\infty} a_h \int_a^b \phi_h(x) g(x) \, dx, \tag{14}$$

reveals an interesting relationship between an L_2-function $f(x)$ and *its Fourier series*

$$\sum_{h=1}^{\infty} a_h \phi_h(x), \quad a_h = \int_a^b \phi_h(x) f(x) \, dx$$

with respect to a complete ON-system $\{\phi_h\}$. It is known that in general this relationship is *not* an equality. It is often indicated by Hurwitz's sign:

$$f(x) \sim \sum_{h=1}^{\infty} a_h \phi_h(x). \tag{14'}$$

Nevertheless, (14′) becomes a true equality if both sides are multiplied by another L_2-function $g(x)$ and then integrated over (a, b) (the case $g(x) \equiv 1$ naturally being included), provided Parseval's equation holds for $f(x)$ and $g(x)$.†

In particular, if $g(x)$ is the step-function

$$g(x) = \begin{cases} 1 & (a \leqslant x \leqslant \xi < b), \\ 0 & (\xi < x \leqslant b), \end{cases} \tag{15}$$

† Or, we can also say, provided the system $\{\phi_h\}$ is complete *with respect to the functions f and g.*

we obtain
$$\int_a^\xi f(x)\,dx = \sum_{h=1}^\infty a_h \int_a^\xi \phi_h(x)\,dx, \tag{16}$$

and it is not difficult to see that the series on the right-hand side converges *uniformly* with respect to ξ.

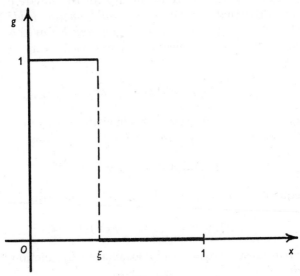

Fig. 6

3·3. The Riesz-Fischer Theorem

Given an ON-system $\{\phi_h(x)\}$ and a sequence of constants

$$a_1, a_2, a_3, \ldots, \tag{1}$$

the convergence of the infinite series Σa_h^2 is evidently a *necessary* condition for the existence of an L_2-function $f(x)$ whose Fourier coefficients with respect to the system $\{\phi_h\}$ are the given constants (1); but, is this condition also *sufficient*?

The famous *Riesz-Fischer theorem* (1907) answers this question unconditionally in the affirmative. It was one of the first brilliant successes of the concept of the Lebesgue integral.†

† Lebesgue integration plays an essential rôle here because the function $f(x)$ corresponding to a given sequence (1) with $\Sigma a_h^2 < \infty$ may not be Riemann integrable.

The proof is very easy if we assume without proof† the (non-trivial!) *Weyl lemma*, which is an analogue, for convergence in the mean, of the classical Cauchy convergence principle. This lemma can be stated as follows:

A necessary and sufficient condition for convergence in the mean over the interval (a, b) of a sequence of L_2-functions

$$f_1(x), \quad f_2(x), \quad f_3(x), \quad \ldots \tag{2}$$

to a certain function $f(x)$ of the same class L_2, is that to any positive ϵ there corresponds a positive integer n_0 such that for any pair m, n of integers both larger than n_0, we have

$$\int_a^b [f_m(x) - f_n(x)]^2 \, dx < \epsilon. \tag{3}$$

Now, assuming this lemma and the convergence of Σa_h^2, we set

$$f_n(x) = a_1 \phi_1(x) + a_2 \phi_2(x) + \ldots + a_n \phi_n(x). \tag{4}$$

If $m > n$, an easy calculation gives us

$$\int_a^b [f_m(x) - f_n(x)]^2 \, dx = \int_a^b \left[\sum_{h=n+1}^{m} a_h \phi_h(x) \right]^2 dx = \sum_{h=n+1}^{m} a_h^2;$$

and, since the series Σa_h^2 is convergent, for any positive ϵ there exists an integer n_0 such that for $n > n_0$

$$\sum_{h=n+1}^{\infty} a_h^2 < \epsilon.$$

Hence for $m > n > n_0$ we have *a fortiori*

$$\int_a^b [f_m(x) - f_n(x)]^2 \, dx < \epsilon,$$

and, by virtue of Weyl's lemma, we can state that

$$\underset{n \to \infty}{\text{l.i.m.}} f_n(x) = f(x), \tag{5}$$

where $f(x)$ is an L_2-function. This function has the given constants (1) as its Fourier coefficients. In fact, if we call $c_1, c_2, c_3, \ldots$ the

---

† For this see, for example, Tricomi [47], p. 33 or J. L. Walsh, *Interpolation and approximation, etc.*, Amer. Math. Soc. Coll. Publ. no. 20, New York, 1935, p. 116.

Fourier coefficients of $f(x)$, the calculation used in the previous section (except for the interchange of a_h and c_h) shows that

$$\int_a^b [f(x) - f_n(x)]^2 \, dx = \int_a^b f^2(x) \, dx - \sum_{h=1}^n c_h^2 + \sum_{h=1}^n (c_h - a_h)^2 \geqslant \sum_{h=1}^n (c_h - a_h)^2.$$

Consequently, if merely one c_h were different from the corresponding a_h, for instance, if $c_1 \neq a_1$, we would have

$$\int_a^b [f(x) - f_n(x)]^2 \, dx \geqslant (c_1 - a_1)^2,$$

in contradiction to (5).

Thus the Riesz-Fischer theorem is proved. If we assume further that the system $\{\phi_h\}$ is *complete*, we can prove that *the function $f(x)$ is essentially unique*. This cannot be deduced directly from (5) because there might exist a completely different function $f^*(x)$ (with the same Fourier coefficients (1)) obtainable in another way as the limit in the mean of the sum (4). However, if the system $\{\phi_h\}$ is complete, this is impossible, since, from Parseval's equation (3·2-10) and the fact that the Fourier coefficients of $f(x) - f^*(x)$ are all zero, it follows that

$$\int_a^b [f(x) - f^*(x)]^2 \, dx = 0. \tag{6}$$

3·4. Completeness and Closure

DEFINITION. *A system of L_2-functions (not necessarily orthogonal) in the range (a, b) will be called closed, in the whole class L_2 or in a subset M of it, if the only functions of class M orthogonal to all the functions of the given system are those which vanish almost everywhere in (a, b).*

Using this definition, we can restate the last result of the previous section as follows:

Any ON-system of functions which is complete in a given set M of L_2-functions is also closed in that set.

We can also prove the converse:

Any ON-system of functions closed in a given class M of L_2-functions is also complete there.

For this we first observe that *any* ON-system $\{\phi_h\}$ is closed in the special class of functions Φ consisting of the functions ϕ_h, their

linear combinations, and the limits in the mean of these combinations:
$$f(x) = \text{l.i.m.}_{n \to \infty} [a_1 \phi_1(x) + a_2 \phi_2(x) + \dots + a_n \phi_n(x)].$$

(If the series Σa_h^2 converges, the existence of the latter function is assured by the reasoning in the previous section.)

In fact, by this reasoning we have

$$\int_a^b f(x)\,\phi_1(x)\,dx = a_1, \quad \int_a^b f(x)\,\phi_2(x)\,dx = a_2, \quad \dots,$$

and consequently these integrals can all vanish only in the trivial case $a_1 = a_2 = \dots = 0$.

Now let the ON-system $\{\phi_h\}$ be closed in the class M and hence (obviously) in the class $M + \Phi$. It is easy to show that any function $g(x)$ of class M can be approximated as closely as desired by means of linear combinations of ϕ_h-functions, i.e. that *the system is complete in* M. To be precise, if we neglect functions vanishing almost everywhere, we can write

$$g(x) = \text{l.i.m.}_{n \to \infty} \sum_{h=1}^{n} a_h \phi_h(x) \quad \text{with} \quad a_h = \int_a^b g(x)\,\phi_h(x)\,dx; \tag{1}$$

because, if
$$g^*(x) = \text{l.i.m.}_{n \to \infty} \sum_{h=1}^{n} a_h \phi_h(x),$$

then $g(x) - g^*(x)$ is a non-trivial function of class $M + \Phi$ whose Fourier coefficients are all *zero*, i.e. a function of class $M + \Phi$ orthogonal to all the ϕ_h-functions and this contradicts our hypothesis.

Hence there is a complete equivalence between completeness and closure for every ON-system of functions.†

But now the reader may ask: Are there any complete systems of functions? (Of course, complete beyond the obvious class Φ.)

We can answer this question easily by considering the step-functions $g(x)$ defined in (3·2-15). Let $r_1, r_2, r_3, \dots$ be the *rational points* of the segment (a, b) ordered in any fashion,‡ and let us put

$$\psi_h(x) = g_{r_h}(x) = \begin{cases} 1 & (a \leqslant x \leqslant r_h < b), \\ 0 & (r_h < x \leqslant b). \end{cases} \tag{2}$$

---

† Even if the system is *not* orthogonal, this equivalence is not destroyed provided completeness is defined by approximation in the mean instead of by Parseval's equation. This explains why some authors use the words *complete* and *closed* with interchanged roles from those used in this book.

‡ Note that the set of rational points of any interval (a, b) is *everywhere dense* (i.e. every point of the interval is an accumulation point of the set) but, in spite of this, it is still *denumerable*.

It is easy to see that the ON-system $\{\phi_h(x)\}$ obtained by ortho-gonalizing (see § 3·1) the ψ_h-functions is *closed in the entire class* L_2.

In fact, any L_2-function $f(x)$ orthogonal to all the ϕ_h must be orthogonal to all the ψ_h as well, i.e.

$$\int_a^{r_h} f(x)\, dx = 0 \quad (h = 1, 2, 3, \ldots).$$

In other words, the *integral function* $F(\xi)$ corresponding to $f(x)$

$$F(\xi) = \int_a^{\xi} f(x)\, dx$$

vanishes at all rational points of the segment (a, b). But such an integral function is continuous (even absolutely continuous); hence $F(\xi) \equiv 0$ everywhere, and $f(x)$ must vanish almost everywhere.†

We shall now formulate some tests for the closure-completeness of an ON-system of L_2-functions, which are easier to use than the bare definitions. We first state an obvious criterion:

I. *No finite ON-system* $\{\phi_h(x)\}$ $(h = 1, 2, \ldots, n)$ *of functions can be closed except in the class* Φ *of linear combinations of the functions* ϕ_h.

In fact, if $\psi(x)$ is any L_2-function outside of Φ, we can easily construct (see § 3·1) a non-null function

$$k_{n+1,\,1}\phi_1(x) + k_{n+1,\,2}\phi_2(x) + \ldots + k_{n+1,\,n}\phi_n(x) + \psi(x),$$

which is orthogonal to all the functions $\phi_1(x)$, $\phi_2(x)$, ..., $\phi_n(x)$ of the system.

Another simple but useful criterion for completeness is:

II. *If the functions of a subset M of the class L_2 (M may even equal L_2) can be approximated (in the mean) as closely as desired by linear combinations of the functions of another subset M' of the class L_2, each system $\{\phi_h\}$ complete in M' is also complete in M.*

In fact, by repeated application of the observation made in extending Parseval's equation to the product of two functions (§ 3·2), we see first that the system $\{\phi_h\}$ remains complete if the subset M' is enlarged to the (generally) more ample subset $M^* \equiv M' + M''$ by adding the set M'' of all linear combinations of any finite number of functions of M'. Secondly, given any function $f(x)$ of M and any positive number ϵ, we can certainly find in M^* a suitable linear combination $f_m^*(x)$ of m functions of M' and a

---

† For $f(x)$ coincides almost everywhere with the derivative of $F(x)$.

suitable linear combination $f_n(x)$ of the first n functions of the system $\{\phi_h\}$ such that†

$$\int_a^b [f(x)-f_m^*(x)]^2\,dx < \tfrac{1}{4}\epsilon, \quad \int_a^b [f_m^*(x)-f_n(x)]^2\,dx < \tfrac{1}{4}\epsilon.$$

But

$$\int_a^b [f(x)-f_n(x)]^2\,dx$$
$$\leqslant 2\int_a^b [f(x)-f_m^*(x)]^2\,dx + 2\int_a^b [f_m^*(x)-f_n(x)]^2\,dx < \epsilon;$$

hence the system $\{\phi_h\}$ is also complete in M.

An immediate consequence of the previous theorem is the so-called *Lauricella criterion* (1912) relative to the case $M \equiv L_2$, which —not only for historical reasons—is worth stating explicitly:

III. *If a certain system $\{\phi_n\}$ of L_2-functions (not necessarily orthonormal) is complete relative to all functions of another system $\{\psi_n\}$, orthogonal or not orthogonal, which is known to be complete in L_2, then the system $\{\phi_h\}$ is also complete in L_2.*

From Lauricella's criterion follows the useful *closure-condition of* Vitali (1921):

IV. *A necessary and sufficient condition for the completeness of an ON-system of L_2-functions $\{\phi_h\}$ in the entire class L_2 is that*

$$\sum_{h=1}^\infty \left[\int_a^\xi \phi_h(x)\,dx \right]^2 = \xi - a \quad (a \leqslant \xi \leqslant b). \tag{3}$$

This is a consequence of III and the fact that the previous system (2) of step-functions is closed (and hence complete) in the entire L_2-class, since (3), for $\xi = r_h$, is precisely Parseval's equation for the function (2) with respect to the system $\{\phi\}$. We thus see that it is sufficient for (3) to hold for only the *rational* values of ξ in (a, b); this is obvious since both sides of (3) are continuous functions of ξ.

If we use only the definition, for the proof of the completeness of an ON-system in the L_2-space it is necessary to verify Parseval's equation for *all* L_2-functions. Now, with the help of Vitali's condition, we see that, as a matter of fact, it is sufficient to consider only the ∞^1 step-functions $g_\xi(x)$.

---

† The first inequality is true because the functions of M can be approximated arbitrarily closely by linear combinations of the M'-functions and the second because the system $\{\phi_h\}$ is complete in M'.

A further improvement is possible, in the sense that, instead of having to consider a series *of functions* like series (3), we merely need to verify that a certain series *of constants* has a certain sum. This was shown by D. P. Dalzell,† who observed that, in view of Bessel's inequality for the step-function $g_\xi(x)$, the difference

$$\Delta(\xi) = \xi - a - \sum_{h=1}^{\infty} \left[\int_a^\xi \phi_h(x)\,dx \right]^2$$

is never negative. Consequently, the condition

$$\int_a^b \Delta(\xi)\,d\xi = 0 \tag{4}$$

is not only *necessary* (this is obvious), but also sufficient for the validity of (3), i.e. for the closure of the system $\{\phi_h\}$. In fact, if the integral of a *non-negative* function vanishes, then the integrand must vanish almost everywhere in the range of integration. Moreover, from (4), we can assert that $\Delta(\xi) \equiv 0$ everywhere, because the integrand is *continuous*. To show this we observe that, if we put

$$\int_a^{\xi_i} \phi_h(x)\,dx = \int_a^b g_{\xi i}(x)\,\phi_h(x)\,dx = a_{hi} \quad (i = 1, 2),$$

we obtain successively, with use of Bessel's inequality,

$$[\Delta(\xi_1) - \Delta(\xi_2) - (\xi_1 - \xi_2)]^2$$

$$= \left[\sum_{h=1}^{\infty} (a_{h1}^2 - a_{h2}^2) \right]^2 = \left[\sum_{h=1}^{\infty} (a_{h1} + a_{h2})(a_{h1} - a_{h2}) \right]^2$$

$$\leqslant \sum_{h=1}^{\infty} (a_{h1} + a_{h2})^2 \sum_{h=1}^{\infty} (a_{h1} - a_{h2})^2$$

$$\leqslant 2 \sum_{h=1}^{\infty} (a_{h1}^2 + a_{h2}^2) \sum_{h=1}^{\infty} \left\{ \int_a^b [g_{\xi_1}(x) - g_{\xi_2}(x)] \phi_h(x)\,dx \right\}^2$$

$$\leqslant 2 \left[\int_a^b g_{\xi_1}^2(x)\,dx + \int_a^b g_{\xi_2}^2(x)\,dx \right] \int_a^b [g_{\xi_1}(x) - g_{\xi_2}(x)]^2\,dx$$

$$\leqslant 4(b-a) \left| \int_{\xi_1}^{\xi_2} dx \right| = 4(b-a)\,|\xi_1 - \xi_2|.$$

It follows that

$$|\Delta(\xi_1) - \Delta(\xi_2)| \leqslant |\xi_1 - \xi_2| + 2\sqrt{[(b-a)\,|\xi_1 - \xi_2|]}.$$

† D. P. Dalzell, *J. London Math. Soc.* **20**, 1945, pp. 87–93.

Another consequence of the continuity of $\Delta(\xi)$, and of a theorem of Dini on the uniform convergence of a series of non-negative functions whose sum is continuous,[†] is the possibility of calculating the integral (4) using term-by-term integration. Here Dalzell's condition assumes the form:

V. *A necessary and sufficient condition for the completeness of an orthonormal system of L_2-functions $\{\phi_h\}$ in the entire class L_2 is that*

$$\sum_{h=1}^{\infty} \int_a^b \left[\int_a^\xi \phi_h(x)\,dx \right]^2 d\xi = \tfrac{1}{2}(b-a)^2. \tag{5}$$

More generally, instead of (4), we can write $\displaystyle\int_a^b \Delta(\xi)\,q(\xi)\,d\xi = 0$,
where $q(\xi)$ is any *non-negative* L_2-function which vanishes at most on a set of measure *zero*; correspondingly, we have the condition

$$\sum_{h=1}^{\infty} \int_a^b \left[\int_a^\xi \phi_h(x)\,dx \right]^2 q(\xi)\,d\xi = \frac{1}{2} \int_a^\xi (\xi-a)\,q(\xi)\,d\xi, \tag{6}$$

which is sometimes easier to verify than (5). For further criterion for completeness see F. G. Tricomi, 'Sulla chiusura dei sistemi ortogonali di funzioni'. *Rev. Un. mat. argent.*, **17**, 299–303 (1955).

3·5. Completeness of the Trigonometric System and of the Polynomials

A very important orthonormal system is the *trigonometric system*:

$$(2\pi)^{-\frac{1}{2}},\quad \pi^{-\frac{1}{2}}\cos x,\quad \pi^{-\frac{1}{2}}\sin x,\quad \pi^{-\frac{1}{2}}\cos 2x,\quad \pi^{-\frac{1}{2}}\sin 2x,\quad \ldots, \tag{1}$$

which, thus written, is orthonormal in the basic interval $(-\pi, \pi)$.

The trigonometric system is complete in the entire class L_2.

We can prove this basic fact by showing that Dalzell's condition (3·4-4) holds for the relative $\Delta(\xi)$ function, namely:

$$\Delta(\xi) = \xi + \pi - (2\pi)^{-1} \left(\int_{-\pi}^\xi dx \right)^2$$

$$- \pi^{-1} \sum_{n=1}^{\infty} \left\{ \left[\int_{-\pi}^\xi \cos nx\,dx \right]^2 + \left[\int_{-\pi}^\xi \sin nx\,dx \right]^2 \right\}$$

$$= \xi + \pi - (2\pi)^{-1}(\xi + \pi)^2 - \pi^{-1} \sum_{n=1}^{\infty} n^{-2} \{ (\sin n\xi)^2 + [\cos n\xi - (-1)^n]^2 \}$$

$$= \tfrac{1}{2}\pi - 2\pi^{-1} \sum_{n=1}^{\infty} n^{-2} - (2\pi)^{-1}\xi^2 + 2\pi^{-1} \sum_{n=1}^{\infty} (-1)^n\, n^{-2} \cos n\xi.$$

[†] U. Dini, *Lezioni di Analisi Infinitesimale*, Nistri, Pisa, 1907–9, **1**, p.e.

We have only to prove that

$$0 = \int_{-\pi}^{\pi} \Delta(\xi)\,d\xi$$

$$= \pi^2 - 4\sum_{n=1}^{\infty} n^{-2} - (2\pi)^{-1}\frac{2\pi^3}{3} + 2\pi^{-1}\sum_{n=1}^{\infty}(-1)^n\, n^{-3}\,[\sin n\xi]_{-\pi}^{\pi}$$

$$= \frac{2\pi^2}{3} - 4\sum_{n=1}^{\infty} n^{-2},$$

i.e. that $$\sum_{n=1}^{\infty} n^{-2} = \tfrac{1}{6}\pi^2. \tag{2}$$

Equation (2), however, is well known.†

We shall now prove that the system of the *polynomials* in any finite interval (a, b) is complete, i.e. that *the (non-orthogonal) system*

$$1, \quad x, \quad x^2, \quad x^3, \quad \dots \tag{3}$$

is closed, and hence complete, in the whole class L^2.

We shall use a method which is based on the observation that, if a function $f(x)$ is orthogonal to all functions (3), it must also be orthogonal to every linear combination of them, i.e. to every *polynomial* and, in particular, to the special polynomial of degree $2n$.

$$P_n(x) = \left[1 + \frac{\delta^2 - (x - x_0)^2}{(b-a)^2}\right]^n,$$

where x_0 and δ are two constants which we shall soon specify.

We shall prove that

$$I_n = \int_a^b P_n(x) f(x)\,dx = 0 \tag{4}$$

is impossible unless $f(x) \equiv 0$, first for the case in which $f(x)$ is continuous and then for $f(x)$ any L_2-function.

If $f(x)$ is any continuous function, not vanishing identically in the interval (a, b), then there must be at least one *interior* point x_0 for which $f(x_0) \neq 0$. Without loss of generality we can assume that $f(x_0) = \eta > 0$. Consequently, the positive number δ can be chosen so small that the interval $(x_0 - \delta,\ x_0 + \delta)$ is entirely contained in (a, b), and

$$f(x) > \tfrac{1}{2}\eta > 0 \quad (x_0 - \delta \leqslant x \leqslant x_0 + \delta). \tag{5}$$

† See, for example, Knopp [25], where there are many proofs of this important equality. See also § 3·10.

On the other hand, in the same interval, we have

$$0 \leqslant \frac{\delta^2 - (x - x_0)^2}{(b-a)^2} < 1,$$

and in a smaller interval $(x_0 - \delta' \leqslant x \leqslant x_0 + \delta')$ $(0 < \delta' < \delta)$ we even have

$$0 < \mu \leqslant \frac{\delta^2 - (x - x_0)^2}{(b-a)^2} < 1,$$

where μ is a suitable positive constant. Hence,

$$(1+\mu)^n \leqslant P_n < 2^n \quad (x_0 - \delta' \leqslant x \leqslant x_0 + \delta');$$

consequently, we have

$$\int_{x_0-\delta}^{x_0+\delta} P_n(x)f(x)\,dx > \int_{x_0-\delta'}^{x_0+\delta'} P_n(x)f(x)\,dx > \tfrac{1}{2}\eta 2\delta'(1+\mu)^n = \eta\delta'(1+\mu)^n.$$

$$(6)$$

Now we consider the remaining parts $(a \leqslant x \leqslant x_0 - \delta)$ and $(x_0 + \delta \leqslant x \leqslant b)$ of the interval of integration, where

$$0 < 1 + \frac{\delta^2 - (x - x_0)^2}{(b-a)^2} \leqslant 1.$$

If we call A an upper bound of the absolute value of the continuous function $f(x)$, we have

$$\left| \int_a^{x_0-\delta} + \int_{x_0+\delta}^b f(x)\,P_n(x)\,dx \right| \leqslant \int_a^{x_0-\delta} + \int_{x_0+\delta}^b |f(x)|\,dx < A(b-a).$$

This shows that this part of the integral I_n remains *bounded* as $n \to \infty$. But, according to (6), the integral over $(x_0 - \delta,\ x_0 + \delta)$ approaches $+\infty$ as $n \to \infty$; hence

$$\lim_{n\to\infty} I_n = \infty,$$

which contradicts (4).

Thus, the assertion has been proved for $f(x)$ continuous. If $f(x)$ is not continuous, but L-integrable,† we set

$$\int_a^\xi f(\xi)\,d\xi = F(\xi).$$

---

† Hence we are not using completely the hypothesis that $f(x) \in L_2$; it is sufficient to suppose $f(x) \in L$.

From (4), where $P_n(x)$ now designates *any* polynomial, we obtain by an integration by parts

$$\int_a^b P_n'(x)\, F(x)\, dx = 0, \tag{7}$$

since obviously $F(a) = 0$ and $F(b) = 0$ because the function $f(x)$ must be orthogonal to the function 1, i.e.

$$\int_a^b f(x)\, dx = 0.$$

But the derivative of a polynomial is a polynomial; hence by the first part of the proof, the *continuous* function $F(x)$ which satisfies condition (7) is identically *zero*, and $f(x)$ must vanish almost everywhere in (a, b).

We have thus proved that *the system* (3) *is complete not only in L_2, but even in L*.

Note that for the results of this section, the hypothesis that the basic interval be finite is essential.

If we attempt to extend the representation of an 'arbitrary' function $f(x)$ in terms of functions of (1) or (3) to an infinite interval, the infinite *series* are no longer useful and instead we have to use infinite integrals, i.e. the *Fourier Integral* (or the *Fourier Transformation*) and the *Mellin* or *Laplace Transformation*.

3·6. Approximation of a General L_2-Kernel by Means of PG-Kernels

Now we shall consider two applications of the previous theory to integral equations. First, we shall prove a theorem which played an essential role in the previous chapter (§ 2·4):

Any L_2-kernel $K(x, y)$, i.e. any kernel for which both functions

$$A(x) = \left[\int_0^1 K^2(x, y)\, dy\right]^{\frac{1}{2}}, \quad B(y) = \left[\int_0^1 K^2(x, y)\, dx\right]^{\frac{1}{2}} \tag{1}$$

exist almost everywhere in the basic interval $(0, 1)$ and belong to the class L_2, can be decomposed (in an infinite number of ways) into the sum of a suitable PG-kernel

$$S(x, y) = \sum_{k=1}^n X_k(x)\, Y_k(y), \tag{2}$$

and another L_2-kernel $T(x, y)$ whose norm can be made smaller than any prescribed positive number ϵ

$$\| T \|^2 = \int_0^1 \int_0^1 T^2(x, y) \, dx \, dy < \epsilon^2. \tag{3}$$

For the proof, we consider any ON-system $\{\phi_h\}$ of L_2-functions complete in the basic interval $(0, 1)$, e.g. the system of orthogonal polynomials obtained by orthogonalizing (see § 3·1) the system $(3·5\text{-}3)$.† Let

$$\int_0^1 K(x, y) \, \phi_h(x) \, dx = a_h(y) \quad (h = 1, 2, 3, \ldots), \tag{4}$$

and observe that the (obviously measurable) functions $a_h(y)$ belong to the class L_2, because from Parseval's equation it follows that

$$\sum_{h=1}^\infty a_h^2(y) = \int_0^1 K^2(x, y) \, dx = B^2(y),$$

and consequently

$$\int_0^1 a_h^2(y) \, dy \leqslant \int_0^1 \left[\sum_{h=1}^\infty a_h^2(y) \right] dy = \int_0^1 B^2(y) \, dy = \| K \|^2.$$

Moreover, from $(3·2\text{-}9)$ and the completeness of the system $\{\phi_h\}$ it follows that

$$\int_0^1 \left[K(x, y) - \sum_{h=1}^n \phi_h(x) \, a_h(y) \right]^2 dx = \sum_{h=n+1}^\infty a_h^2(y).$$

The series on the right-hand side can be integrated term by term by virtue of Lebesgue's fundamental theorem, since its terms are all non-negative; hence we obtain

$$\int_0^1 \int_0^1 \left[K(x, y) - \sum_{h=1}^n \phi_h(x) \, a_h(y) \right]^2 dx \, dy = \sum_{h=n+1}^\infty \int_0^1 a_h^2(y) \, dy.$$

If we put

$$\sum_{h=1}^n \phi_h(x) \, a_h(y) = S(x, y), \quad K(x, y) - S(x, y) = T(x, y), \tag{5}$$

and if we call R_n the sum of the terms following the nth in the convergent series of constants:

$$\sum_{h=1}^\infty \int_0^1 a_h^2(y) \, dy = \int_0^1 B^2(y) \, dy = \| K \|^2,$$

---

† These polynomials are very similar to *Legendre polynomials*, which arise from orthogonalizing the powers of x in the basic interval $(-1, 1)$.

then we can write

$$\| T \|^2 = \int_0^1 \int_0^1 T^2(x, y)\, dx\, dy = R_n. \tag{6}$$

This shows that inequality (3) is certainly satisfied, provided that n is so large that $R_n < \epsilon^2$.

Finally, we observe that $S(x, y)$ is a true PG-kernel; for, if the functions $a_1(y)$, $a_2(y)$, ..., $a_n(y)$ are linearly dependent, then the number n of terms of the sum may be reduced so that the new functions of y become linearly independent.

3·7. Enskog's Method

Another interesting application of the previous theory to Fredholm integral equations with a *general* kernel (i.e. one not necessarily symmetric) is Enskog's method (1926) for the numerical solution of Fredholm integral equations of the second kind

$$\phi(x) - \lambda \int_0^1 K(x, y)\, \phi(y)\, dy = f(x). \tag{1}$$

The basis of this method is Green's formula (2·1-15) which together with (1) gives us the equality

$$\int_0^1 \mathscr{F}_x^*[\Phi(y)]\, \phi(x)\, dx = \int_0^1 f(x)\, \Phi(x)\, dx, \tag{2}$$

where $\Phi(x)$ is an 'arbitrary' function for which Green's formula holds. In particular, we can set $\Phi = \phi_h$ $(h = 1, 2, ...)$, where $\{\phi_h(x)\}$ is *any* fixed ON-system complete in $(0, 1)$ whose individual functions permit the application of Green's formula, e.g. the orthogonal polynomials of the previous section. Then setting

$$\mathscr{F}_x^*[\phi_h(y)] = \psi_h(x) \tag{3}$$

and

$$\int_0^1 f(x)\, \phi_h(x)\, dx = a_h, \tag{4}$$

we deduce from (2) that

$$\int_0^1 \phi(x)\, \psi_h(x)\, dx = a_h. \tag{5}$$

In other words, we can calculate the 'Fourier coefficients'

of the unknown function $\phi(x)$ with respect to the system of functions

$$\psi_1(x) = \mathscr{F}_x^*[\phi_1(y)], \quad \psi_2(x) = \mathscr{F}_x^*[\phi_2(y)], \quad \dots \quad (6)$$

The difficulty is that in general this system is *not* orthogonal.

However, if the functions $\{\psi_h\}$ are linearly independent, we can *orthonormalize* system (6) (see § 3·1), thus obtaining a complete ON-system of the form

$$\left.\begin{aligned}
\chi_1(x) &= \alpha_{11}\psi_1(x), \\
\chi_2(x) &= \alpha_{21}\psi_1(x) + \alpha_{22}\psi_2(x), \\
\chi_3(x) &= \alpha_{31}\psi_1(x) + \alpha_{32}\psi_2(x) + \alpha_{33}\psi_3(x), \\
&\dots\dots\dots\dots\dots\dots\dots\dots\dots\dots\dots
\end{aligned}\right\} \quad (7)$$

With respect to this system, the Fourier coefficients of the unknown function are

$$\left.\begin{aligned}
b_1 &= \alpha_{11}a_1, \\
b_2 &= \alpha_{21}a_1 + \alpha_{22}a_2, \\
b_3 &= \alpha_{31}a_1 + \alpha_{32}a_2 + \alpha_{33}a_3, \\
&\dots\dots\dots\dots\dots\dots\dots\dots
\end{aligned}\right\} \quad (8)$$

Consequently the unknown function $\phi(x)$ can be calculated 'theoretically' by the formula

$$\phi(x) = \underset{n\to\infty}{\text{l.i.m.}} \sum_{h=1}^{n} b_h \chi_h(x), \quad (9)$$

and practically by its identification with one of the sums on the right-hand side corresponding to a sufficiently large n.

The method is useless if and only if the system $\{\psi_h\}$ cannot be orthogonalized. This, of course, can only happen if it is a linearly dependent system, i.e. if and only if there is an integer n and n constants $\mu_1, \mu_2, \dots, \mu_n$ (not all zero) such that

$$0 \equiv \mu_1\psi_1(x) + \dots + \mu_n\psi_n(x) = \mathscr{F}_x^*[\mu_1\phi_1(y) + \dots + \mu_n\phi_n(y)].$$

The function under the operator $\mathscr{F}^*$ is certainly not zero almost everywhere because the system $\{\phi_h\}$ is linearly independent (since it is orthogonal); hence Enskog's method is useless if and only if there exists a non-null function $\Phi(x)$ such that

$$0 \equiv \mathscr{F}_x^*[\Phi(y)] \equiv \Phi(x) - \lambda \int_0^1 K(y, x)\, \Phi(y)\, dy,$$

i.e. *if and only if* λ *is an eigenvalue*† of the given integral equation.

† Remember that $K(x, y)$ and $K(y, x)$ have the same eigenvalues (§ 2·3).

Except for this case, both systems $\{\psi_h\}$ and $\{\chi_h\}$ are complete, because, in view of the identities

$$\int_0^1 F(x)\,\psi_h(x)\,dx = \int_0^1 \mathscr{F}_x[F(y)]\,\phi_h(x)\,dx$$

$$= \int_0^1 \left[F(x) - \lambda \int_0^1 K(x,y)\,F(y) \right] \phi_h(x)\,dx$$

$$(h = 1, 2, \ldots),$$

and in view of the completeness of the system $\{\phi_h\}$, there will exist a non-null function $F(x)$ orthogonal to all the ψ_h if and only if this function satisfies the homogeneous equation

$$F(x) - \lambda \int_0^1 K(x,y)\,F(y)\,dy = 0,$$

i.e. *if and only if λ is an eigenvalue.*

Enskog's method, which can also be used to construct independently the theory of Fredholm integral equations, is especially useful if the functions $\phi_h(x)$ are not too different from the eigenfunctions of the symmetric kernel

$$K(x,y) + K(y,x) - \lambda \int_0^1 K(x,z)\,K(y,z)\,dz \tag{10}$$

(which we shall meet later (§ 3·16)), because in this case the functions $\{\psi_h\}$ are 'nearly' orthogonalized.†

3·8. The Spectrum of a Symmetric Kernel

Let us now return to the *symmetric* kernels, and begin by observing that if
$$K(x,y) = K(y,x), \tag{1}$$
we also have $$K_n(x,y) = K_n(y,x) \quad (n = 2, 3, \ldots) \tag{2}$$
and $$H(x,y;\lambda) = H(y,x;\lambda). \tag{3}$$

It is also easy to show that *the eigenvalues and eigenfunctions of any real symmetric kernel are real.*‡

† If the functions ϕ_h coincide *exactly* with these eigenfunctions, then the functions $\{\psi_h\}$ are already orthogonalized. See, for example, Tricomi [48], p. 154.

‡ Since any eigenfunction may be multiplied by an arbitrary constant, which need not be real, we shall mean by the reality of a set of eigenfunctions that these functions can be made real by multiplication by a suitable constant factor or by the formation of suitable linear combinations (in the case of eigenvalues of higher index).

In fact, if a symmetric kernel has a complex eigenvalue $\lambda_1 = a + ib$ with a corresponding eigenfunction $\phi_1(x) = \alpha(x) + i\beta(x)$, then it must also have the conjugate eigenvalue $\lambda_2 = a - ib$ with a corresponding eigenfunction $\phi_2(x) = \alpha(x) - i\beta(x)$. Since the eigenfunctions of a symmetric kernel corresponding to two *different* eigenvalues λ_1 and λ_2 are orthogonal, we have†

$$\int \phi_1(x)\,\phi_2(x)\,dx = \int [\alpha^2(x) + \beta^2(x)]\,dx = 0.$$

This contradicts the hypothesis that ϕ_1 and ϕ_2 are *eigenfunctions*, i.e. functions which do not vanish almost everywhere.

Another important property of symmetric kernels is that *their spectrums are never empty*:‡

Every symmetric, non-null L_2-kernel has at least one eigenvalue.

According to the final observations in § 2·5 about relations between the spectrum of a kernel $K(x, y)$ and that of its iterated kernels, it is sufficient to carry out the proof for the iterated kernel $K_2(x, y)$. To be precise, it is sufficient to show that the infinite series (2·5-18) corresponding to K_2, i.e. the infinite series

$$A_2 + A_4\lambda^2 + A_6\lambda^4 + \dots, \tag{4}$$

where

$$A_{2m} = \int K_{2m}(x, x)\,dx \tag{5}$$

has a *finite* radius ρ of convergence. Moreover, if we can obtain an upper bound ρ^* for this ρ, we shall also obtain a corresponding bound

$$|\lambda_1| = \sqrt{\rho} \leqslant \sqrt{\rho^*} \tag{6}$$

for the minimal absolute value of the eigenvalues.§

For this we observe that from the formula

$$K_{m+n}(x, y) = \int K_m(x, z)\,K_n(z, y)\,dz$$

for *symmetric* kernels it follows that

$$K_{m+n}(x, x) = \int K_m(x, y)\,K_n(x, y)\,dy,$$

---

† As in the previous chapter, we shall not indicate the limits of integration when these limits are $(0, 1)$.

‡ The *spectrum* of a kernel is the set of its eigenvalues.

§ In other words, we call λ_1 the eigenvalue with the minimal absolute value. If $\pm \lambda_1$ are both eigenvalues, then λ_1 will designate the *positive* one.

and consequently, using the Schwarz inequality for double integrals, we have

$$A_{m+n}^2 = \left[\int K_{m+n}(x,x)\,dx \right]^2 = \left[\int\int K_m(x,y)\,K_n(x,y)\,dx\,dy \right]^2$$

$$\leqslant \int\int K_m^2(x,y)\,dx\,dy \int\int K_n^2(x,y)\,dx\,dy$$

$$= \int K_{2m}(x,x)\,dx \int K_{2n}(x,x)\,dx = A_{2m}A_{2n}.$$

In particular, changing m and n into $n-1$ and $n+1$, respectively, we obtain

$$A_{2n}^2 \leqslant A_{2n-2}A_{2n+2} \qquad (n = 2, 3, \ldots). \tag{7}$$

This inequality shows at once that the quantities A_2, A_4, A_6, ... (which are certainly not negative) are definitely *positive*. In fact, if $A_{2n+2} = 0$ then $A_{2n} = 0$, and consequently

$$A_{2n-2} = A_{2n-4} = \ldots = A_4 = 0.$$

However, if

$$A_4 = \int K_4(x,x)\,dx = 0,$$

then, since

$$K_4(x,x) = \int K_2^2(x,y)\,dy \geqslant 0,$$

we see that $K_4(x,x) \equiv 0$ almost everywhere. This shows that, if we neglect at most a null subset of $0 \leqslant x \leqslant 1$, we must also have

$$\int K_2^2(x,y)\,dy = 0.$$

Hence, neglecting at most a null subset of the interval $0 \leqslant x \leqslant 1$ and a similar one of $0 \leqslant y \leqslant 1$, we must have $K_2(x,y) \equiv 0$. This implies that

$$\int K_2(x,x)\,dx = \| K \|^2 = 0,$$

which contradicts our hypothesis. Moreover, we see that $A_2 = \| K \|^2$ is positive.

The fact that $A_{2m} > 0$ $(m = 1, 2, \ldots)$ allows us to put inequality (7) in the form

$$\frac{A_{2n}}{A_{2n+2}} \leqslant \frac{A_{2n-2}}{A_{2n}};$$

hence the *positive* quotient A_{2n}/A_{2n+2} does not increase with increasing n and consequently approaches a *finite* limit $L > 0$ as

$n \to \infty$. Since this limit coincides with the radius ρ of convergence of the series (4), we have

$$\rho = \lim_{n \to \infty} \frac{A_{2n}}{A_{2n+2}}, \tag{8}$$

and also

$$\lambda_1^2 \leqslant \frac{A_{2n}}{A_{2n+2}} \quad (n = 1, 2, 3, \ldots). \tag{9}$$

In particular, for $n = 1$ we obtain the very useful upper *bound*

$$|\lambda_1| \leqslant \sqrt{\frac{A_2}{A_4}} = \frac{\|K\|}{\|K_2\|}. \tag{10}$$

We have thus proved that each symmetric (non-zero) kernel $K(x, y)$ has at least one eigenvalue λ_1 with a corresponding eigenfunction $\phi_1(x)$, which we can assume normalized. It follows that if the symmetric kernel

$$K^{(2)}(x, y) = K(x, y) - \frac{\phi_1(x) \phi_1(y)}{\lambda_1}$$

is not zero, it will have at least one eigenvalue λ_2 with corresponding normalized eigenfunction $\phi_2(x)$. Moreover, even though λ_1 may equal λ_2, we can be certain that $\phi_2(x) \neq \phi_1(x)$, for $\phi_1(x)$ cannot be an eigenfunction of the 'shortened' kernel $K^{(2)}(x, y)$, since

$$\int K^{(2)}(x, y) \phi_1(y)\, dy = \int K(x, y) \phi_1(y)\, dy - \frac{\phi_1(x)}{\lambda_1} \int \phi_1^2(y)\, dy \equiv 0.$$

Continuing in this way, we consider the third symmetric kernel

$$K^{(3)}(x, y) = K^{(2)}(x, y) - \frac{\phi_2(x) \phi_2(y)}{\lambda_2} = K(x, y) - \sum_{h=1}^{2} \frac{\phi_h(x) \phi_h(y)}{\lambda_h}$$

and so forth.

There are two possibilities:

Either the process stops after n steps, i.e. $K^{(n+1)} \equiv 0$ and the kernel is a Pincherle-Goursat kernel

$$K(x, y) = \sum_{h=1}^{n} \frac{\phi_h(x) \phi_h(y)}{\lambda_h}, \tag{11}$$

or the process can be continued indefinitely and then there are *an infinite number of eigenvalues*.† Hence:

Every non-zero symmetric kernel either has an infinite number of eigenvalues or is a PG-kernel.

---

† There are effectively an infinite number of different eigenvalues because each value can appear in the sequence $\lambda_1, \lambda_2, \lambda_3, \ldots$ only a *finite* number of times, i.e. the number of times indicated by its *index*.

In either case, we shall assume in future discussions that the spectrum of the kernel is arranged so that

$$|\lambda_1| \leqslant |\lambda_2| \leqslant |\lambda_3| \leqslant ..., \tag{12}$$

with precedence given to the positive eigenvalues over the negative ones (in case $\pm \lambda_i$ are both eigenvalues). Furthermore, we shall suppose that, if there are some eigenvalues of an index $r > 1$, they are repeated r times and the corresponding r linearly independent eigenfunctions are orthogonalized (see § 3·1) and then normalized, like the remaining eigenfunctions.

Therefore if we call

$$\phi_1(x), \quad \phi_2(x), \quad \phi_3(x), \quad ... \tag{13}$$

the eigenfunctions corresponding to $\lambda_1, \lambda_2, \lambda_3, ...$, respectively, they will form an orthonormal system of functions in the basic interval $(0, 1)$, i.e.

$$(\phi_h, \phi_k) = \begin{cases} 0 & (h \neq k), \\ 1 & (h = k), \end{cases} \tag{14}$$

for all h, k.

3·9. The Bilinear Formula

Unfortunately, the *bilinear formula* (3·8-11), valid in the case of a finite number of eigenvalues, cannot be extended to the general case, because (as examples can show) the infinite series corresponding to the sum on the right-hand side of (3·8-11) need not converge. We can, however, write

$$K(x, y) \sim \sum_{h=1}^{\infty} \frac{\phi_h(x) \, \phi_h(y)}{\lambda_h} \tag{1}$$

in the sense of the earlier formula (3·2-14'), because the Fourier coefficients of $K(x,y)$ considered as a function of y (or of x) with respect to the ON-system (3·8-13) are obviously $\phi_h(x)/\lambda_h$ (or $\phi_h(y)/\lambda_h$) since for eigenfunctions

$$\int K(x,y) \, \phi_h(y) \, dy = \frac{\phi_h(x)}{\lambda_h} \quad (h = 1, 2, 3, ...). \tag{2}$$

Later we shall point out some cases (other than the case of a finite number of eigenvalues) in which Hurwitz's sign $\sim$ in (1) can be replaced by the equality sign. Meanwhile we shall show that even if system $\{\phi_h\}$ is incomplete† (which happens quite often), *series* (1) *converges in the mean to the kernel* $K(x, y)$, i.e. we shall show that

$$K(x, y) = \text{l.i.m.} \sum_{h=1}^{n} \frac{\phi_h(x)\,\phi_h(y)}{\lambda_h}, \qquad (3)$$

in the sense that

$$\lim_{n\to\infty} \iint \left[K(x, y) - \sum_{h=1}^{n} \lambda_h^{-1}\,\phi_h(x)\,\phi_h(y) \right]^2 dx\,dy = 0.$$

That the series must converge in the mean in one dimension (i.e. with respect to x or to y only) is a consequence of the Riesz-Fischer theorem (§ 3·3) and Bessel's inequality, since for almost every y in $(0,\ 1)$, using the notation of §§ 2·2 and 3·7, we have

$$\sum_{h=1}^{\infty} [\phi_h(x)/\lambda_h]^2 \leqslant \int K^2(x, y)\,dy = A^2(x) = B^2(x).$$

It is not obvious, however, that the symmetric function of x and y

$$K^*(x, y) = \text{l.i.m.} \sum_{h=1}^{n} \frac{\phi_h(x)\,\phi_h(y)}{\lambda_h} \qquad (4)$$

must coincide almost everywhere with $K(x, y)$, even if the system is incomplete.

We shall prove this by showing that the symmetric kernel

$$R(x, y) = K(x, y) - K^*(x, y) \qquad (5)$$

has no eigenvalues nor eigenfunctions and consequently must vanish almost everywhere in the square $(0 \leqslant x \leqslant 1,\ 0 \leqslant y \leqslant 1)$.

(i) We note first that

$$\int R(x, y)\,\phi_h(y)\,dy = 0 \quad (h = 1, 2, 3, \ldots), \qquad (6)$$

since, as we saw in the proof of the Riesz-Fischer theorem, the function $K^*(x, y)$ has the same Fourier coefficients $\phi_h(y)/\lambda_h$ as $K(x, y)$ with respect to the ON-system $\{\phi_h(x)\}$.

(ii) Using (5) we show secondly that any eigenfunction $\psi(x)$ of $R(x, y)$, i.e. any normalized function for which

$$\psi(x) = \mu \int R(x, y)\,\psi(y)\,dy, \qquad (7)$$

† Otherwise the fact would be trivial.

where μ is a suitable constant, must be orthogonal to all the functions of the system $\{\phi_h\}$, i.e.

$$(\psi, \phi_h) = 0 \quad (h = 1, 2, 3, \ldots). \tag{8}$$

In fact, using (6) and (7), we have

$$\int \psi(x)\,\phi_h(x)\,dx = \mu \int \phi_h(x)\,dx \int R(x,y)\,\psi(y)\,dy$$

$$= \mu \int \psi(y)\,dy \int R(x,y)\,\phi_h(x)\,dx = 0.$$

(iii) Thirdly, the function ψ must also satisfy the condition

$$I \equiv \int K^*(x,y)\,\psi(y)\,dy = 0. \tag{9}$$

Indeed, using (8) and the notation

$$\sum_{h=1}^{n} \frac{\phi_h(x)\,\phi_h(y)}{\lambda_h} = \Phi_n(x,y), \tag{10}$$

we have for any positive integer n

$$I = \int [K^*(x,y) - \Phi_n(x,y)]\,\psi(y)\,dy + \sum_{h=1}^{n} \frac{\phi_h(x)}{\lambda_h} \int \phi_h(y)\,\psi(y)\,dy$$

$$= \int [K^*(x,y) - \Phi_n(x,y)]\,\psi(y)\,dy.$$

Consequently if n is large enough, and if we use the Schwarz inequality, we obtain from (4)

$$I^2 \leqslant \int [K^*(x,y) - \Phi_n(x,y)]^2\,dy \int \psi^2(y)\,dy$$

$$= \int [K^*(x,y) - \Phi_n(x,y)]^2\,dy < \epsilon,$$

ϵ having the usual significance. Since I is independent of n, this implies that $I = 0$.

With these preparations, we can prove immediately that the eigenfunction $\psi(x)$ does not exist. In fact, from (7) and (9) it follows that

$$\mu \int K(x,y)\,\psi(y)\,dy = \mu \int R(x,y)\,\psi(y)\,dy + \mu \int K^*(x,y)\,\psi(y)\,dy = \psi(x).$$

Hence $\psi(x)$, which is orthogonal to all the functions $\{\phi_h\}$, is also an eigenfunction of the given kernel $K(x,y)$, i.e. it is one of the functions $\phi_h(x)$ and so must be orthogonal to itself.

Using the previous result we can prove:

If the infinite series
$$\sum_{h=1}^{\infty} \frac{\phi_h(x)\,\phi_h(y)}{\lambda_h} \tag{11}$$

converges uniformly, or at least so that to any positive ϵ there corresponds a positive integer n_0 such that

$$\iint \left[\sum_{h=n+1}^{\infty} \frac{\phi_h(x)\,\phi_h(y)}{\lambda_h}\right]^2 dx\,dy < \epsilon \quad (n > n_0), \tag{12}$$

then
$$K(x,y) = \sum_{h=1}^{\infty} \frac{\phi_h(x)\,\phi_h(y)}{\lambda_h}. \tag{13}$$

If we call $S(x,y)$ the sum of series (11), we have from (12)

$$\iint [S(x,y) - \Phi_n(x,y)]^2\,dx\,dy < \epsilon,$$

where Φ_n is given by (10). From (3) we obtain

$$\iint [K(x,y) - \Phi_n(x,y)]^2\,dx\,dy < \epsilon,$$

where n is larger than some n_0^*. Consequently, if

$$n > \max(n_0, n_0^*),$$

we have

$$\iint [S(x,y) - K(x,y)]^2\,dx\,dy \leqslant 2\iint [S(x,y) - \Phi_n(x,y)]^2\,dx\,dy$$
$$+ 2\iint [K(x,y) - \Phi_n(x,y)]^2\,dx\,dy$$
$$< 4\epsilon.$$

Thus $S(x,y) \equiv K(x,y)$ almost everywhere in the square

$$(0 \leqslant x \leqslant 1,\ 0 \leqslant y \leqslant 1).$$

Another important consequence of (3) is the following necessary and sufficient condition for the closure of a system of eigenfunctions:

Any L_2-function $w(x)$ is orthogonal to all the eigenfunctions $\phi_h(x)$ of the symmetric kernel $K(x,y)$ if and only if

$$\int K(x,y)\,w(y)\,dy = 0, \tag{14}$$

i.e. *if and only if the function w is 'orthogonal to the kernel $K(x, y)$' itself.*

In fact, we have identically

$$\int \phi_h(x) \, dx \int K(x, y) \, w(y) \, dy = \int w(y) \, dy \int K(x, y) \, \phi_h(x) \, dx$$

$$= \lambda_h^{-1} \int w(y) \, \phi_h(y) \, dy \quad (h = 1, 2, \ldots). \quad (15)$$

Thus, if (14) holds, it follows that

$$\int w(x) \, \phi_h(x) \, dx = 0 \quad (h = 1, 2, \ldots). \quad (16)$$

Conversely, if conditions (16) are satisfied, for any positive integer n we find

$$I = \int K(x, y) \, w(y) \, dy = \int [K(x, y) - \Phi_n(x, y)] \, w(y) \, dy;$$

consequently

$$I^2 \leqslant \int [K(x, y) - \Phi_n(x, y)]^2 \, dy \int w^2(y) \, dy.$$

Denoting the last integral by W and using (3), we have for n large enough

$$I^2 \leqslant W\epsilon$$

(with the usual meaning for ϵ). Hence $I = 0$, and the theorem is proved.

3·10. The Hilbert-Schmidt Theorem and Its Applications

Using the results of the previous section, we can now easily prove the following important theorem:

HILBERT-SCHMIDT THEOREM. *If $f(x)$ can be written in the form*

$$f(x) = \int K(x, y) \, g(y) \, dy, \quad (1)$$

where $K(x, y)$ is a symmetric L_2-kernel and $g(y)$ is an L_2-function, then $f(x)$ can also be represented by its Fourier series with respect to the ON-system $\{\phi_h\}$ of eigenfunctions of $K(x, y)$;† i.e. we can write

$$f(x) = \sum_{h=1}^{\infty} a_h \phi_h(x), \quad (2)$$

† For this condition (1) is *sufficient*, but *not necessary*. The question is connected with the Fredholm equations of the *first kind*, which we shall consider later (§ 3·15).

where
$$a_h = \int f(x)\,\phi_h(x)\,dx \quad (h=1,2,3,\dots). \tag{3}$$

Moreover, if $K \in L_2^*$, i.e. if the function $A(x)$ related to the kernel K is bounded:

$$\int K^2(x,y)\,dy = A^2(x) \leqslant N^2 \quad (N = \text{constant}), \tag{4}$$

then series (2) *converges absolutely and uniformly for every $f(x)$ of type* (1). *If $K \in L_2$ only, then series* (2) *converges almost uniformly in the sense of § 2·1.*†

For the proof of the first part of the theorem it is sufficient to observe that from (3·9-15) it follows that

$$
\begin{aligned}
b_h &\equiv \int g(x)\,\phi_h(x)\,dx \\
&= \lambda_h \int f(x)\,\phi_h(x)\,dx = a_h \lambda_h \quad (h=1,2,\dots),
\end{aligned} \tag{5}
$$

and consequently, for any positive integer n, with the same notation as (3·9-10), we have

$$
\begin{aligned}
f(x) &= \int K(x,y)\,g(y)\,dy \\
&= \int [K(x,y) - \Phi_n(x,y)]\,g(y)\,dy + \sum_{h=1}^{n} \frac{\phi_h(x)}{\lambda_h} \int \phi_h(y)\,g(y)\,dy \\
&= \int [K(x,y) - \Phi_n(x,y)]\,g(y)\,dy + \sum_{h=1}^{n} a_h \phi_h(x).
\end{aligned}
$$

Hence, using the Schwarz inequality, we can write

$$\left[f(x) - \sum_{h=1}^{n} a_h \phi_h(x) \right]^2 \leqslant \int [K(x,y) - \Phi_n(x,y)]^2\,dy \int g^2(y)\,dy,$$

and, in view of (3·9-3), we see that

$$
\begin{aligned}
\lim_{n\to\infty} &\left[f(x) - \sum_{h=1}^{n} a_h \phi_h(x) \right]^2 \\
&\leqslant \int_0^1 g^2(y)\,dy \lim_{n\to\infty} \int [K(x,y) - \Phi_n(x,y)]^2\,dy = 0,
\end{aligned}
$$

† Remember that this signifies that the partial sums admit an upper bound which is an L-function; the main consequence is that the series may be integrated term by term. (Lebesgue's fundamental theorem.)

i.e. we obtain equality (2). Furthermore, if our kernel is an L_2^*-*kernel,* and if we use the Schwarz inequality for sums,† then for any positive integer n, we obtain

$$\left[\sum_{h=n+1}^{\infty} |a_h \phi_h(x)|\right]^2 = \left[\sum_{h=n+1}^{\infty} \left| b_h \frac{\phi_h(x)}{\lambda_h}\right|^2\right] \leqslant \sum_{h=n+1}^{\infty} \frac{\phi_h^2(x)}{\lambda_h^2} \sum_{h=n+1}^{\infty} b_h^2,$$

and *a fortiori*

$$\sum_{h=n+1}^{\infty} |a_h \phi_h(x)| \leqslant N\left[\sum_{h=n+1}^{\infty} b_h^2\right]^{\frac{1}{2}}, \tag{6}$$

because from (3·9-2), the Bessel inequality (3·2-8) and (4), it follows that

$$\sum_{h=n+1}^{\infty} \frac{\phi_h^2(x)}{\lambda_h^2} \leqslant \int K^2(x, y)\, dy = A^2(x) \leqslant N^2.$$

Since $g(x)$ belongs to the class L_2 the series Σb_h^2 is convergent. Hence, given any positive ϵ, if n is larger than a suitable n_0 *independent of* x, we shall have

$$\sum_{h=n+1}^{\infty} b_h^2 < \epsilon^2 \quad (n > n_0),$$

and from (6) it follows that

$$\sum_{h=n+1}^{\infty} |a_h \phi_h(x)| < N\epsilon.$$

This shows that series (2) converges *absolutely and uniformly.*

It remains to show that if the kernel $K(x, y)$ belongs only to the class L_2, then series (2) converges *almost uniformly,* i.e. that its partial sums admit an upper bound which is a summable function in the sense of Lebesgue. This is clear because, without using (4), we have

$$\left[\sum_{h=1}^{n} |a_h \phi_h(x)|\right]^2 \leqslant \sum_{h=1}^{n} \frac{\phi_h^2(x)}{\lambda_h^2} \sum_{h=1}^{n} b_h^2$$

$$\leqslant A^2(x)\int g^2(x)\, dx = A^2(x)\|g\|^2.$$

† We are referring to the well-known inequality (*Inequalities* [15], p. 16)

$$\left(\sum_{h=1}^{n} a_h b_h\right)^2 \leqslant \sum_{h=1}^{n} a_h^2 \sum_{h=1}^{n} b_h^2,$$

which is sometimes called *Cauchy's inequality.* It is valid even for $n = \infty$, provided both series on the right converge.

The Hilbert-Schmidt theorem can be easily memorized by observing that the relation with Hurwitz's sign

$$g(y) \sim \sum_{h=1}^{\infty} b_h \phi_h(y)$$

(which generally is *not* an equality) becomes an equality (with an absolutely and almost uniformly convergent series on the right) if we multiply both sides by $K(x, y)$ and then integrate over $(0, 1)$, because

$$\int K(x, y) g(y) \, dy = \sum_{h=1}^{\infty} b_h \int K(x, y) \phi_h(y) \, dy = \sum_{h=1}^{n} \frac{b_h}{\lambda_h} \phi_h(x)$$

is just (2). It should be pointed out that the possibility of this transformation is not an immediate consequence of the similar observation of § 3·2, because the system of eigenfunctions generally is *not* complete.

The Hilbert-Schmidt theorem is important not only in the theory of integral equations, but also in the theory of orthogonal functions. Among its consequences is the important fact that the bilinear formula (3·9-1) is valid *with the equality sign* for all the *iterated* kernels $K_m(x, y)$ $(m = 2, 3, \ldots)$. This is obvious because, by definition,

$$K_m(x, y) = \int K(x, z) K_{m-1}(z, y) \, dz \quad (m = 2, 3, \ldots),$$

and this equation is of the form (1) with $g \equiv K_{m-1}$. The Fourier coefficient $a_n(y)$ of $K_m(x, y)$ (considered as a function of x) with respect to the system of eigenfunctions of $K(x, y)$ is

$$a_h(y) = \int K_m(x, y) \phi_h(x) \, dx = \lambda_h^{-m} \phi_h(y) \quad (h = 1, 2, 3, \ldots),$$

since (§ 2·5) $\phi_h(x)$ is an eigenfunction of K_m corresponding to the eigenvalue λ_h^m. Hence:

If the symmetric kernel $K(x, y)$ belongs to the class L_2, then all the corresponding iterated kernels $K_m(x, y)$ $(m \geqslant 2)$ can be represented by the absolutely and almost uniformly convergent series

$$K_m(x, y) = \sum_{h=1}^{\infty} \lambda_h^{-m} \phi_h(x) \phi_h(y) \quad (m = 2, 3, \ldots). \tag{7}$$

If the kernel $K(x, y)$ belongs to the class L_2^*, i.e. if condition (4) is satisfied, then each series converges uniformly.

In particular, putting $y = x$, integrating between 0 and 1, and remembering (2·5-17), we obtain

$$\sum_{h=1}^{\infty} \lambda_h^{-m} = A_m \quad (m = 2, 3, \ldots), \tag{8}$$

where $A_1, A_2, \ldots$ are the *traces* of the given kernel $K(x, y)$. As we shall see in § 3·11, this formula is *very important* in practical applications of the theory of integral equations.

From (8) we know that the eigenvalues of any L_2-kernel are such that the infinite series $\Sigma \lambda_h^{-2}$ converges. Moreover, this is a *unique* 'asymptotic' property of such eigenvalues, because, given any sequence of real numbers $\lambda_1, \lambda_2, \lambda_3, \ldots$ satisfying this condition, and any ON-system $\{\phi_h(x)\}$ of *uniformly bounded L_2-functions*,† the Riesz-Fischer theorem insures the convergence in the mean of the series

$$\sum_{h=1}^{\infty} \frac{\phi_h(x) \, \phi_h(y)}{\lambda_h}$$

to a symmetric L_2-function $K(x, y)$ which, considered as a kernel of a Fredholm integral equation, has precisely the numbers $\lambda_1, \lambda_2, \lambda_3, \ldots$ as eigenvalues.

We shall now make use of the Hilbert-Schmidt theorem to find an explicit solution of the non-homogeneous Fredholm equation of the second kind with a symmetric kernel

$$\phi(x) - \lambda \int K(x, y) \, \phi(y) \, dy = f(x), \tag{9}$$

where λ is not an eigenvalue.

For this it is sufficient to note that, since, according to (9), the function $\phi(x) - f(x)$ has an integral representation of the form (1), it can also be represented by an absolutely and almost uniformly convergent series of the type

$$\phi(x) - f(x) = \sum_{h=1}^{\infty} c_h \phi_h(x),$$

where $\qquad c_h = \int [\phi(x) - f(x)] \, \phi_h(x) \, dx = \xi_h - a_h$

with $\qquad \xi_h = \int \phi(x) \, \phi_h(x) \, dx, \quad a_h = \int f(x) \, \phi_h(x) \, dx.$

---

† This means that $|\phi_h(x)| < M$, where M is a positive number independent of h and x. For instance, the ON-system (15) below satisfies this condition.

From (5) it follows that $\quad c_h = \lambda \dfrac{\xi_h}{\lambda_h}$.

Hence, since $\lambda \neq \lambda_h$ $(h = 1, 2, \ldots)$, we have

$$\xi_h = \frac{\lambda_h}{\lambda_h - \lambda} a_h, \quad c_h = \frac{\lambda}{\lambda_h - \lambda} a_h,$$

and we obtain the solution of our equation in terms of the absolutely and almost uniformly convergent series

$$\phi(x) = f(x) + \lambda \sum_{h=1}^{\infty} \frac{a_h}{\lambda_h - \lambda} \phi_h(x), \tag{10}$$

which can also be written

$$\phi(x) = f(x) - \lambda \sum_{h=1}^{\infty} \int \frac{\phi_h(x)\,\phi_h(y)}{\lambda - \lambda_h} f(y)\, dy. \tag{11}$$

We see also that the resolvent kernel $H(x, y; \lambda)$ can be expressed by the elegant series

$$H(x, y; \lambda) = \sum_{h=1}^{\infty} \frac{\phi_h(x)\,\phi_h(y)}{\lambda - \lambda_h}, \tag{12}$$

if this series converges almost uniformly in the basic interval $(0, 1)$, because, in such a case, the integral and summation signs in (11) may be interchanged.

Independently from this last condition, equation (2·1-10) for the resolvent kernel together with the Hilbert-Schmidt theorem gives us the absolutely and almost uniformly convergent expansion

$$H(x, y; \lambda) = -K(x, y) + \lambda \sum_{h=1}^{\infty} \frac{\phi_h(x)\,\phi_h(y)}{\lambda_h(\lambda - \lambda_h)}. \tag{12′}$$

This expansion shows that *the singular points of the resolvent kernel H corresponding to a symmetric L_2-kernel $K(x, y)$ are simple poles.*

The Hilbert-Schmidt theorem can also be used to prove that twice differentiable functions can be expanded in Fourier series, in the usual sense of the word, i.e. in convergent trigonometric series. To be precise, putting the statement in a form which allows us to avoid a few difficulties of detail, we can prove the following theorem:

If the function $f(x)$ can be differentiated twice in the basic interval $(0, 1)$, and if its second derivative $f''(x)$ belongs to the class L_2, and if

$f(0) = f(1) = 0$, *then $f(x)$ has the absolutely and uniformly convergent expansion*

$$f(x) = \sum_{h=1}^{\infty} \alpha_h \sin (h\pi x) \quad (0 \leqslant x \leqslant 1), \tag{13}$$

where

$$\alpha_h = 2\int_0^1 f(x) \sin (h\pi x) \, dx \quad (h = 1, 2, 3, \dots). \tag{14}$$

If we note that $\alpha_h/\sqrt{2}$ is the Fourier coefficient of the given function $f(x)$ with respect to the ON-system of functions

$$\sqrt{2}\sin (\pi x), \quad \sqrt{2}\sin (2\pi x), \quad \sqrt{2}\sin (3\pi x), \quad \dots, \tag{15}$$

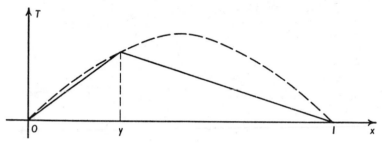

Fig. 7

the statement can be considered as an immediate consequence of the Hilbert-Schmidt theorem, provided we can form an L_2^*-kernel $K(x, y)$ with the eigenfunctions (15), such that any function satisfying the above conditions can be expressed by means of an integral of type (1).

This is accomplished by using the *triangular kernel*:

$$T(x, y) = \begin{cases} (1-x) y & (0 \leqslant y \leqslant x \leqslant 1), \\ x(1-y) & (0 \leqslant x \leqslant y \leqslant 1), \end{cases} \tag{16}$$

which belongs to the class L_2^* (it is even continuous). It is pictured in Fig. 7, where the dotted curve is the parabola representing the function

$$T(x, x) = x(1-x).$$

Since

$$\int_0^1 T(x, y)\,\phi(y)\,dy = (1-x)\int_0^x y\phi(y)\,dy + x\int_x^1 (1-y)\,\phi(y)\,dy$$

$$= \int_0^x y\phi(y)\,dy + x\int_x^1 \phi(y)\,dy - x\int_0^1 y\phi(y)\,dy$$

holds identically, it follows that

$$\frac{d}{dx} \int_0^1 T(x,y)\,\phi(y)\,dy = \int_x^1 \phi(y)\,dy - \int_0^1 y\phi(y)\,dy$$

and that
$$\frac{d^2}{dx^2} \int_0^1 T(x,y)\,\phi(y)\,dy = -\phi(x). \tag{17}$$

Hence the homogeneous equation

$$\phi(x) - \lambda \int_0^1 T(x,y)\,\phi(y)\,dy = 0$$

is equivalent to the differential equation

$$\frac{d^2\phi(x)}{dx^2} + \lambda\phi(x) = 0 \tag{18}$$

with the boundary conditions

$$\phi(0) = \phi(1) = 0, \tag{19}$$

which arise from the fact that $T(0,y) \equiv T(1,y) \equiv 0$. The only solutions of (18) satisfying the first of conditions (19) are of the form $C\sin(\sqrt{\lambda}\,x)$, and these satisfy the second condition of (19) if and only if
$$\sqrt{\lambda} = h\pi,$$

where h is any integer; hence the eigenvalues of $T(x,y)$ are

$$\lambda_1 = \pi^2, \quad \lambda_2 = (2\pi)^2, \quad \lambda_3 = (3\pi)^3, \quad \ldots,$$

and the corresponding (normalized) eigenfunctions are, as prescribed, the functions (15).

Moreover,
$$T(x,y) = 2 \sum_{h=1}^{\infty} \frac{\sin(h\pi x)\sin(h\pi y)}{(h\pi)^2}, \tag{20}$$

since the series on the right is obviously uniformly convergent.

Finally, we observe that every function $f(x)$ which satisfies the conditions of the statement can be written in form (1). In fact, from (17) it follows that

$$\frac{d^2}{dx^2} \left[\int_0^1 T(x,y)f''(y)\,dy + f(x) \right] \equiv 0;$$

consequently, with the conditions at $x = 0$ and $x = 1$, we have

$$f(x) = -\int_0^1 T(x,y)f''(y)\,dy. \tag{21}$$

Putting $y = x$ and integrating, we deduce from (20) that

$$\frac{2}{\pi^2} \sum_{h=1}^{\infty} \frac{1}{h^2} \int_0^1 \sin^2(h\pi x)\, dx = \frac{1}{\pi^2} \sum_{h=1}^{\infty} \frac{1}{h^2} = \int_0^1 T(x, x)\, dx$$

$$= \int_0^1 x(1-x)\, dx = \tfrac{1}{6},$$

that is,
$$\sum_{h=1}^{\infty} \frac{1}{h^2} = \frac{\pi^2}{6}, \tag{22}$$

a well-known, but not trivial, result which we used in §3·5 to prove the completeness of the trigonometric system.

3·11. Extremal Properties and Bounds for Eigenvalues

The theory of integral equations with a symmetric kernel is not only closely connected with the theory of simultaneous linear equations, but also with the theory of quadratic forms and the calculus of variations. More exactly, it is connected with the problem of the extrema of Hilbert's double integral

$$J(\phi, \phi) = \iint K(x, y)\, \phi(x)\, \phi(y)\, dx\, dy. \tag{1}$$

This double integral is sometimes called *a quadratic form with an infinite number of variables*, because it can be considered as the limiting case as $n \to \infty$ of the quadratic form

$$\sum_{r,\, s=1}^{n} k_{rs} x_r x_s, \tag{2}$$

where
$$k_{rs} = k_{sr} = \frac{1}{n^2} K\left(\frac{r}{n}, \frac{s}{n}\right), \quad x_r = \phi\left(\frac{r}{n}\right) \quad (r, s = 1, 2, \ldots, n).$$

This leads to many ideas, for instance, that of finding a counterpart to the reduction of a quadratic form to its *canonical form* $\Sigma \alpha_r x_r^2$. The counterpart is *Hilbert's formula*

$$J(\phi, \phi) = \sum_{h=1}^{\infty} \frac{a_h^2}{\lambda_h}, \quad a_h = \int \phi_h(x)\, \phi(x)\, dx, \tag{3}$$

where the λ_h and the $\phi_h(x)$ are the eigenvalues and eigenfunctions of the L_2-kernel $K(x, y)$.

In fact, by virtue of the Hilbert-Schmidt theorem we have

$$\int K(x, y)\, \phi(y)\, dy = \sum_{h=1}^{\infty} \frac{a_h}{\lambda_h} \phi_h(x),$$

where the series on the right converges absolutely and almost uniformly. Multiplying by $\phi(x)$ and integrating between 0 and 1, we obtain (3) without further ado. Similarly, if we consider the *bilinear form*

$$J(\phi, \psi) \equiv \int\int K(x, y)\,\phi(x)\,\psi(y)\,dx\,dy,$$

we have

$$J(\phi, \psi) = J(\psi, \phi) = \sum_{h=1}^{\infty} \frac{a_h b_h}{\lambda_h},$$

where

$$a_h = \int \phi(x)\,\phi_h(x)\,dx, \quad b_h = \int \psi(x)\,\phi_h(x)\,dx.$$

If we identify the arbitrary L_2-function $\phi(x)$ with the general eigenfunction $\phi_m(x)$, we obtain from (3)

$$J(\phi_m, \phi_m) = \frac{1}{\lambda_m} \quad (m = 1, 2, 3, \ldots), \tag{4}$$

and, in particular,

$$J(\phi_1, \phi_1) = \frac{1}{\lambda_1}. \tag{4'}$$

On the other hand, since

$$\frac{1}{|\lambda_1|} \geqslant \frac{1}{|\lambda_2|} \geqslant \frac{1}{|\lambda_3|} \geqslant \ldots,$$

it follows from (3) and Bessel's inequality that

$$J(\phi, \phi) \leqslant \sum_{h=1}^{\infty} \frac{a_h^2}{|\lambda_h|} \leqslant \frac{1}{|\lambda_1|} \sum_{h=1}^{\infty} a_h^2 \leqslant \frac{1}{|\lambda_1|} \int \phi^2(x)\,dx,$$

that is,

$$|\lambda_1| \leqslant \|\phi\|^2 / J(\phi, \phi), \tag{5}$$

and, in particular,

$$|\lambda_1| \leqslant [J(\phi, \phi)]^{-1}, \tag{6}$$

provided that $\phi(x)$ is normalized, i.e. provided that

$$\int \phi^2(x)\,dx = 1.$$

This, together with (4′) shows that $|J(\phi, \phi)|$ *has a maximum* in the space L_N of *normalized* L_2-functions, and that this maximum is attained for $\phi \equiv \phi_1$, and has the value $|\lambda_1|^{-1}$.† In other words, we have obtained the important formula

$$|\lambda_1|^{-1} = \max |J(\phi, \phi)| \quad (\phi \in L_N). \tag{7}$$

† However, $J(\phi, \phi)$ has no *minimum* except the *lower bound* zero, as long as the number of eigenvalues is infinite.

Similarly, if we consider the space $L_{N,m}$ of L_2-functions $\phi(x)$ satisfying the m conditions

$$\int \phi^2(x)\,dx = 1, \quad (\phi, \phi_1) = (\phi, \phi_2) = \ldots = (\phi, \phi_{m-1}) = 0, \qquad (8)$$

together with (4), we have

$$|J(\phi, \phi)| = \left| \sum_{h=m}^{\infty} \frac{a_h^2}{\lambda_h} \right| \leqslant \sum_{h=m}^{\infty} \frac{a_h^2}{|\lambda_h|} \leqslant \frac{1}{|\lambda_m|} \sum_{h=1}^{\infty} a_h^2$$

$$\leqslant \frac{1}{|\lambda_m|} \int \phi^2(x)\,dx = \frac{1}{|\lambda_m|},$$

and consequently we can write

$$|\lambda_m|^{-1} = \max |J(\phi, \phi)| \qquad (\phi \in L_{N,m};\ m = 1, 2, 3, \ldots). \qquad (9)$$

In spite of their simplicity, the previous results are of great importance. For example, we note (following R. Courant)[†] that in many cases the *natural frequencies* $\nu_1, \nu_2, \nu_3, \ldots$ of a mechanical vibrating system (e.g. a vibrating bar, bell, etc.) are (as in § 1·9 for the bar) proportional to the square root of the (positive) eigenvalues $\lambda_1, \lambda_2, \lambda_3, \ldots$ of a Fredholm equation of the second kind, whose unknown function ϕ is generally proportional to the amplitude of the oscillations of a generic point of the system.

Consequently, in view of (9), the mth natural frequency ν_m is given by a formula of the type

$$\nu_m = C_m [\max |J(\phi, \phi)|]^{-\frac{1}{2}}, \qquad (10)$$

where C_m is a constant, and ϕ ranges over a suitable functional space S_m. This space is usually smaller than the previous space $L_{N,m}$, since in addition to conditions (8), some others are generally needed, for instance, the continuity of ϕ, if the vibrating system is not broken. However, if the vibrating system has a *split* along a certain line l, even functions ϕ discontinuous across this line l can be admitted, and the space S_m must be replaced by a larger one S'_m. Then we have

$$\max_{\phi \in S'_m} |J(\phi, \phi)| \geqslant \max_{\phi \in S_m} |J(\phi, \phi)|$$

and

$$\nu'_m \leqslant \nu_m,$$

† See [6], I, Ch. 6.

where ν'_m is the corresponding natural frequency of the 'broken' system. This explains the common experience that a split bell, a split china dish, etc., has a duller clang than an unbroken one! However, if new constraints are added, for instance, if some points of the vibrating system are fixed, then the space S_m must be replaced by a smaller one S''_m, and consequently (with obvious notations) $\nu''_m \geqslant \nu_m$, another common experience!

Another important consequence of the previous results is that

$$|\lambda_m| \leqslant |J(\phi_0, \phi_0)|^{-1}, \tag{11}$$

where ϕ_0 denotes *any* L_2-function satisfying conditions (8) and (possibly) the other conditions imposed by the problem.

In particular,

$$|\lambda_1| \leqslant |J(\phi_0, \phi_0)|^{-1}, \tag{12}$$

where ϕ_0 denotes *any normalized L_2-function*, and this or the equivalent formula (5), in addition to (3·8-10), gives us valuable upper bounds for the first eigenvalue, which is often the most important or the only one.

For instance, in the case of the *triangular kernel* $T(x, y)$, whose first eigenvalue is

$$\lambda_1 = \pi^2 = 9 \cdot 86960,$$

formula (3·8-10) gives us the majorant

$$\lambda_1 < \sqrt{(105)} = 10 \cdot 2470, \tag{13}$$

while formula (12) with $\phi_0 \equiv 1$ gives us

$$\lambda_1 < 12, \tag{14}$$

since $\quad J(1, 1) = \int_0^1 \left[(1-x) \int_0^x y\,dy + x \int_x^1 (1-y)\,dy \right] dx = \frac{1}{12}.$

In this case, as often happens, the approximation given by (3·8-10) is better than that given by (12), as long as we utilize the rough approximation $\phi_0 \equiv 1$ for the first eigenfunction.

The approximation furnished by (12) can be substantially improved by following an idea of Ritz, i.e. by using a normalized function $\phi_0 = \phi_0(x, \xi_1, \xi_2, ..., \xi_n)$ containing a finite number of undetermined parameters $\xi_1, \xi_2, ..., \xi_n$ which can be successively calculated, using the usual theory of maxima and minima of functions of several variables, by means of the condition

$$J(\phi_0, \phi_0) = \max.$$

In the case of the triangular kernel, we can identify ϕ_0 with the normalized step-function $\phi_0(x, \xi)$ represented in Fig. 8. With elementary calculations we find

$$J(\phi_0, \phi_0) = \tfrac{1}{12}(1 + 2\xi - 8\xi^2) \quad (0 \leqslant \xi < \tfrac{1}{2}),$$

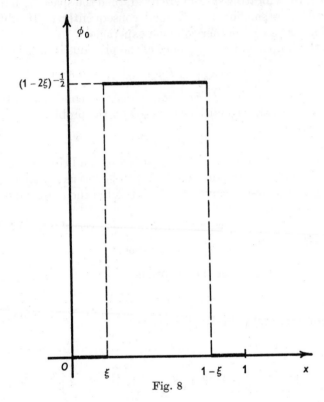

Fig. 8

a function which reaches its maximum value $\tfrac{3}{32}$ at $\xi = \tfrac{1}{8}$. Consequently, we obtain a new upper bound

$$\lambda_1 < \tfrac{32}{3} = 10\cdot666..., \tag{15}$$

which is better than the previous bound (14).

The method of Ritz corresponds precisely to the choice

$$\phi_0 = \xi_1 \psi_1(x) + \xi_2 \psi_2(x) + ... + \xi_n \psi_n(x) \quad (\xi_1^2 + \xi_2^2 + ... + \xi_n^2 = 1),$$

where $\{\psi_h\}$ is *any* ON-system of functions in the basic interval $(0, 1)$. This method gives good numerical results (especially for the calcu-

lation of the *first* eigenvalue λ_1) if the system $\{\psi_h\}$ does not 'differ too much' from the system of eigenfunctions of the kernel $K(x, y)$.†
The calculation of the following eigenvalues is generally more difficult, in spite of the fact that the difficulty arising from the presence of the first $m-1$ eigenfunctions in the accessory conditions (8) can be avoided, as Courant showed.‡

All the previous bounds are *upper* bounds for the eigenvalues. A *lower* bound for the first eigenvalue λ_1 (or for λ_n if $\lambda_1, \lambda_2, ..., \lambda_{n-1}$ are already known), can be easily deduced from the earlier formula (3·10-8).§ For, neglecting all the terms of the series which follow the first (which are usually small), we obtain

$$| \lambda_1 |^{-m} \leqslant | A_m | = \left| \int K_m(x, x)\, dx \right|,$$

that is,
$$| \lambda_1 | \geqslant | A_m |^{-1/m} \quad (m = 2, 3, ...). \tag{16}$$

For instance, with the triangular kernel, since

$$T_m(x, y) = 2\pi^{-2m} \sum_{n=1}^{\infty} n^{-2m} \sin(n\pi x) \sin(n\pi y), \tag{17}$$

we obtain the following lower bounds:

$$\lambda_1 > \left[\pi^{-2m} \sum_{n=1}^{\infty} n^{-2m} \right]^{-1/m} = \frac{1}{4} \left[\frac{(-1)^{m-1}}{2(2m)!} B_{2m} \right]^{-1/m} \quad (m = 1, 2, 3, ...), \tag{18}$$

where the B_{2m} are the *Bernoulli numbers*.¶ In particular, for $m = 1, 2, 3$ we have the successively improved bounds

$$\lambda_1 > 6, \quad \lambda_1 > \sqrt{(90)} = 9{\cdot}4868, \quad \lambda_1 > \sqrt[3]{(945)} = 9{\cdot}8132. \tag{19}$$

As a by-product, we obtain the interesting formula

$$\lim_{m \to \infty} \left[\frac{(-1)^{m-1}}{2(2m)!} B_{2m} \right]^{-1/m} = 4\pi^2. \tag{20}$$

Sometimes the approximation obtained by using (16) can be substantially improved if instead of simply neglecting the terms $\lambda_2^{-m} + \lambda_3^{-m} + ...$ of the series (3·10-8), we substitute approximate

---

† Formula (12) would give the *exact* value of λ_1, if we put $\phi_0(x) \equiv \phi_1(x)$, but the eigenfunction $\phi_1(x)$ is generally unknown.
‡ Courant-Hilbert [6], **1**, p. 352.
§ Apart from the obvious bound, $| \lambda_1 | \geqslant \| K \|^{-1}$.
¶ See, for example, Knopp [25], pp. 182, 239. It is even possible to consider the value $m = 1$ since the kernel T is *positive* (see the following section).

values of $\lambda_2, \lambda_3, \dots$. These could be deduced, for example, from a simplified version of the original problem.†

In any case, a more useful way of obtaining an approximate value for the first eigenvalue is to use the double inequality

$$| A_m |^{-1/m} \leqslant | \lambda_1 | \leqslant \sqrt{\frac{A_{2n}}{A_{2n+2}}} \quad (m = 2, 3, \dots; n = 1, 2, 3, \dots), \quad (21)$$

which is one of the most frequently used results of the theory of integral equations in applied mathematics. This inequality can also be used for the successive eigenvalues, provided we are able to evaluate approximately the first eigenfunction $\phi_1(x)$, which can be used to *remove* the first eigenvalue λ_1 from $K(x, y)$ by constructing the 'shortened' kernel

$$K^{(2)}(x, y) = K(x, y) - \frac{\phi_1(x)\,\phi_1(y)}{\lambda_1}$$

and so forth (cf. § 3·8).

3·12. Positive Kernels—Mercer's Theorem

Hilbert's basic formula

$$J(\phi, \phi) = \sum_{h=1}^{\infty} \frac{a_h^2}{\lambda_h}, \quad a_h = \int \phi(x)\,\phi_h(x)\,dx, \quad (1)$$

together with its corollary

$$J(\phi_h, \phi_h) = \frac{1}{\lambda_h} \quad (h = 1, 2, 3, \dots), \quad (2)$$

shows that *a necessary and sufficient condition for the quadratic form $J(\phi, \phi)$ to be non-negative in L_2-space, i.e. for the inequality*

$$J(\phi, \phi) \geqslant 0 \quad (\phi \in L_2) \quad (3)$$

to hold, is that all the eigenvalues of $K(x, y)$ be positive, i.e.

$$0 < \lambda_1 \leqslant \lambda_2 \leqslant \lambda_3 \leqslant \dots. \quad (4)$$

In such a case we call $K(x, y)$ a *positive* kernel. A kernel is called *positive-definite* if

$$J(\phi, \phi) > 0 \quad (5)$$

for $\| \phi \| > 0$.

The additional condition for *definiteness* is obviously the *closure* of the system of eigenfunctions, i.e. the closure of the kernel itself (§ 3·9).

---

† For an example of this, see Krall-Einaudi [28], **2**, p. 239.

The values of a positive kernel K are not necessarily all positive;† but if $K(x,y)$ is *continuous*, at least in a neighborhood of the diagonal $x=y$ of the basic square $S \equiv (0 \leqslant x \leqslant 1, 0 \leqslant y \leqslant 1)$, then its values on this diagonal must be necessarily *non-negative*, i.e.

$$K(x,x) \geqslant 0 \quad (0 \leqslant x \leqslant 1). \tag{6}$$

In fact, suppose

$$K(x_0, x_0) < 0 \quad (0 < x_0 < 1).$$

(If $x_0 = 0$ or $x_0 = 1$, then the following reasoning must be slightly modified in an obvious manner.) Then, since $K(x,y)$ is assumed to be continuous in a neighborhood of the point $x=y=x_0$, we can find a positive δ so small that the square

$$S_\delta = (x_0 - \delta \leqslant x \leqslant x_0 + \delta, \ x_0 - \delta \leqslant y \leqslant x_0 + \delta)$$

is completely contained in S and, inside of it, $K(x,y) < 0$.

Consequently, if

$$\phi_0(x) = \begin{cases} 1 & (x_0 - \delta \leqslant x \leqslant x_0 + \delta), \\ 0 & (0 \leqslant x < x_0 - \delta, \ x_0 + \delta < x \leqslant 1), \end{cases}$$

then necessarily

$$J(\phi_0, \phi_0) = \iint_{S_\delta} K(x,y)\,dx\,dy < 0.$$

This contradicts the hypothesis that $K(x,y)$ is a positive kernel.

Hence (6) is a *necessary* (but not sufficient) condition for a *continuous* kernel $K(x,y)$ to be positive.

From this condition Mercer's theorem follows:

MERCER'S THEOREM. *If the symmetric L_2-kernel $K(x,y)$ is continuous and has only positive eigenvalues (or at most a finite number of negative eigenvalues), then the series*

$$\Sigma \phi_h(x) \phi_h(y)/\lambda_h$$

converges absolutely and uniformly, and the bilinear formula

$$K(x,y) = \sum_{h=1}^{\infty} \frac{\phi_h(x)\phi_h(y)}{\lambda_h} \tag{7}$$

holds.

† For instance, the values of the symmetric kernel $K(x,y) = f(x)f(y)$ may be partly negative, but the kernel is *positive*, since in this case

$$J(\phi, \phi) = \left[\int f(x)\phi(x)\,dx \right]^2 \geqslant 0.$$

In view of a result in § 3·9, this reduces to the proof of the uniform convergence of series (7). Since the convergence is not influenced by casting aside a finite number of terms, it is no restriction to assume that *all* the eigenvalues are positive, i.e. that $K(x, y)$ is a *positive kernel*. If $K(x, y)$ is positive and continuous, the 'shortened' kernel

$$K^{(m+1)}(x, y) = K(x, y) - \sum_{h=1}^{m} \frac{\phi_h(x)\, \phi_h(y)}{\lambda_h}$$

is also *positive* and continuous for every m.† Therefore, in view of (6), we have

$$K(x, x) - \sum_{h=1}^{m} \frac{\phi_h^2(x)}{\lambda_h} = K^{(m+1)}(x, x) \geqslant 0,$$

that is, $$\sum_{h=1}^{m} \frac{\phi_h^2(x)}{\lambda_h} \leqslant K(x, x) \quad (m = 1, 2, 3, \ldots).$$

This proves that the series with positive terms $\Sigma \phi_h^2(x)/\lambda_h$ is convergent.

Now, for any pair of positive integers n, p, the Schwarz inequality for sums gives

$$\left[\sum_{h=n+1}^{n+p} \left| \frac{\phi_h(x)\, \phi_h(y)}{\lambda_h} \right| \right]^2 \leqslant \sum_{h=n+1}^{n+p} \frac{\phi_h^2(x)}{\lambda_h} \sum_{h=n+1}^{n+p} \frac{\phi_h^2(y)}{\lambda_h}. \tag{7'}$$

Thus for fixed y, the series $\Sigma \phi_h(x)\, \phi_h(y)/\lambda_h$ converges absolutely and uniformly in x and, for a fixed x, absolutely and uniformly in y. This is already sufficient to establish the validity of the bilinear formula (7) and, in particular, of the formula

$$K(x, x) = \Sigma \phi_h^2(x)/\lambda_h.$$

Moreover since the convergence of this last series is uniform in x by virtue of Dini's theorem,‡ we see thus that $\Sigma \phi_h(x)\, \phi_h(y)/\lambda_h$ converges uniformly even when x and y vary simultaneously.

This theorem is very useful because in many mechanical or physical problems, reducible to Fredholm equations with symmetric kernels, the positiveness of the kernels is an immediate consequence of the physical significance of Hilbert's quadratic

---

† Remember that according to the formulas of § 2·5 the eigenfunctions of a continuous kernel are also continuous.

‡ *Dini's theorem*: If the sum of an infinite series of non-negative continuous functions is a continuous function of x in a closed interval $(a \leqslant x \leqslant b)$, then the series converges uniformly there. (See Knopp [25], p. 345.)

form (cf. §3·14). Consequently, in these cases, the bilinear form (7) can be applied without further ado.

COROLLARY. *Under the hypotheses of Mercer's theorem, the infinite series $\Sigma \lambda_h^{-1}$ converges absolutely. We have*

$$\sum_{h=1}^{\infty} \lambda_h^{-1} = A_1 = \int K(x,x)\, dx, \tag{8}$$

and the earlier formula (3·11-16) *can be used even for* $m = 1$.

It should be noticed that the previous formula (3·10-22) may be considered as an application of this corollary to the triangular kernel $T(x, y)$.

Finally, we note that the first iterated kernel $K_2(x, y)$ is always positive since $\lambda_h^2 > 0$. Conversely, any positive, continuous kernel $K(x, y)$ with positive eigenvalues $\lambda_1, \lambda_2, \ldots$ and eigenfunctions $\phi_1(x), \phi_2(x), \ldots$, can be considered as the first iterate of the symmetric kernel

$$K^*(x, y) = \underset{n \to \infty}{\text{l.i.m.}} \sum_{h=1}^{n} \lambda_h^{-\frac{1}{2}} \phi_h(x)\, \phi_h(y)$$

which exists because, as we have just seen, the infinite series

$$\sum_{h=1}^{\infty} [\lambda_h^{-\frac{1}{2}} \phi_h(x)]^2$$

is always convergent.

3·13. Connection with the Theory of Linear Differential Equations

Just as a linear differential equation

$$a_0(x) \frac{d^n y}{dx^n} + a_1(x) \frac{d^{n-1} y}{dx^{n-1}} + \ldots + a_n(x)\, y = F(x), \tag{1}$$

with the classical *initial* conditions

$$y(0) = c_0, \quad y'(0) = c_1, \quad \ldots, \quad y^{(n-1)}(0) = c_{n-1}, \tag{2}$$

leads to a Volterra integral equation (§ 1·8), the same equation with accessory conditions at *both* the end-points $x = 0$ and $x = 1$ of the basic interval (0, 1) leads generally to an integral equation of the Fredholm type.

A very general class of such accessory conditions is obtained by requiring that n linear combinations of the $2n$ quantities

$$y(0), \quad y'(0), \quad \ldots, \quad y^{(n-1)}(0); \quad y(1), \quad y'(1), \quad \ldots, \quad y^{(n-1)}(1)$$

assume prescribed values, i.e. by requiring that

$$\sum_{k=0}^{n-1} a_{hk} y^{(k)}(0) - \sum_{k=0}^{n-1} b_{hk} y^{(k)}(1) = c_h \quad (h = 0, 1, \ldots, n-1), \tag{3}$$

where a_{hk}, b_{hk} and c_h are given constants. Obviously we exclude the case where all the a_{hk} and b_{hk} vanish for the same h. We can suppose, however, that $c_0 = c_1 = \ldots = c_{n-1} = 0$, i.e. we can consider without loss of generality the *homogeneous* boundary conditions

$$\sum_{k=0}^{n-1} a_{hk} y^{(k)}(0) = \sum_{k=0}^{n-1} b_{hk} y^{(k)}(1) \quad (h = 0, 1, 2, \ldots, n-1). \tag{4}$$

This is clear, because if $y(x)$ is a solution of (1) satisfying boundary conditions (3), and $f_0(x)$ is *any* function (differentiable n times) satisfying (3), and if $y(x) = f_0(x) + Y(x)$, then $Y(x)$ satisfies homogeneous conditions (4) and differential equation (1) with only the right-hand side (generally) changed.

Equation (1) plus boundary conditions (4) is called a *Sturm-Liouville system*. Such a system is equivalent to a Fredholm integral equation of the second kind with a rather manageable kernel. We shall confine ourselves here to the most important case, $n = 2$. In this case a certain formal difficulty disappears. For $n > 2$, see books on differential equations, e.g. Ince [21], Chs. x–xi.

The difficulty which disappears for $n = 2$ is the necessity of considering together with equation (1) its *adjoint*,[†] because a linear differential equation of the *second order* can always be made self-adjoint.[‡] To be precise, under the usual hypothesis that $a_0(x) \neq 0$, this can be done by multiplying equation (1) (in the case $n = 2$) by $p(x)/a_0(x)$, where

$$p(x) = \exp \int [a_1(x)/a_0(x)] \, dx > 0. \tag{5}$$

We suppose further that $F(x) \equiv 0$ (which is no essential restriction)[§] and we thus obtain the *self-adjoint equation*

$$\frac{d}{dx}\left[p(x) \frac{dy}{dx} \right] + q(x) y = 0, \tag{6}$$

---

† See, for example, Ince [21], p. 123. However, knowledge of the *adjoint* is not needed here.

‡ Ince [21], p. 215.

§ The solution of a non-homogeneous linear differential equation can be deduced from those of the corresponding homogeneous equation by quadratures. See also formula (30) below.

where
$$q(x) = p(x) \frac{a_2(x)}{a_0(x)}. \tag{7}$$

Equation (6) is called self-adjoint because if we denote by $\mathscr{L}[y(x)]$ the differential operator

$$\mathscr{L}[y(x)] \equiv \frac{d}{dx}\left[p(x)\frac{dy(x)}{dx}\right] + q(x)\,y(x) \tag{8}$$

(i.e. the left-hand side of (6)), then we have identically

$$z(x)\,\mathscr{L}[y(x)] - y(x)\,\mathscr{L}[z(x)] = \frac{d}{dx}\left\{p(x)\left[z(x)\frac{dy(x)}{dx} - y(x)\frac{dz(x)}{dx}\right]\right\}. \tag{9}$$

(In the case of a non-self-adjoint equation a similar identity requires the use of *two* differential operators $\mathscr{L}$ and, say, $\mathscr{M}$ on the left.)

Boundary conditions (4) can be written in the simple form

$$\mathscr{A}_h[y(0)] = \mathscr{B}_h[y(1)] \quad (h = 0, 1) \tag{10}$$

if we introduce the differential operators

$$\mathscr{A}_h[y(x)] = a_{h0}y(x) + a_{h1}y'(x), \quad \mathscr{B}_h[y(x)] = b_{h0}y(x) + b_{h1}y'(x)$$
$$(h = 0, 1)$$

and the abbreviations

$$\mathscr{A}_h[y(x_0)] \equiv \{\mathscr{A}_h[y(x)]\}_{x=x_0}, \quad \mathscr{B}_h[y(x_0)] \equiv \{\mathscr{B}_h[y(x)]\}_{x=x_0}.$$

Generally the Sturm-Liouville system (6) + (10) has only the trivial solution $y(x) \equiv 0$ because, if $y_1(x)$, $y_2(x)$ is a *fundamental set* of solutions of (6) and we write its general solution in the form

$$y(x) = C_1 y_1(x) + C_2 y_2(x),$$

then the determination of the constants C_1, C_2 by means of the boundary conditions (10) leads to the homogeneous algebraic system

$$\begin{cases} C_1\{\mathscr{A}_0[y_1(0)] - \mathscr{B}_0[y_1(1)]\} + C_2\{\mathscr{A}_0[y_2(0)] - \mathscr{B}_0[y_2(1)]\} = 0, \\ C_1\{\mathscr{A}_1[y_1(0)] - \mathscr{B}_1[y_1(1)]\} + C_2\{\mathscr{A}_1[y_2(0)] - \mathscr{B}_1[y_2(1)]\} = 0, \end{cases}$$

which *generally* has only the solution $C_1 = C_2 = 0$, because *in general*

$$\Delta = \begin{vmatrix} \mathscr{A}_0[y_1(0)] - \mathscr{B}_0[y_1(1)] & \mathscr{A}_0[y_2(0)] - \mathscr{B}_0[y_2(1)] \\ \mathscr{A}_1[y_1(0)] - \mathscr{B}_1[y_1(1)] & \mathscr{A}_1[y_2(0)] - \mathscr{B}_1[y_2(1)] \end{vmatrix} \neq 0. \tag{11}$$

We shall, in fact, assume in the following discussion that $\Delta \neq 0$.

Instead of considering the system (6) + (10), we shall consider a slightly different system consisting of boundary conditions (10) and a modified differential equation (containing a parameter λ):

$$\mathscr{L}[y(x)] + \lambda r(x)\,y(x) = \frac{d}{dx}\left[p(x)\frac{dy}{dx}\right] + [q(x) + \lambda r(x)]\,y = 0. \quad (12)$$

We shall ask: *For what values of the parameter does the Sturm-Liouville system* (12) + (10) *have non-trivial solutions?*

The study of this question, which is of fundamental importance in many problems of mathematical physics, mechanics, etc., can be reduced to a study of a homogeneous Fredholm integral equation of the second kind. To do this we shall use *Green's formula* (9) and *Green's function* for the system (6) + (10).

Green's function $G(x, \xi)$ is a function of x and a parameter $\xi\,(0 < \xi < 1)$ satisfying three conditions:

(i) $G(x, \xi)$, considered as a function of x, satisfies the differential equation (6) at all points of the interval (0, 1) except at the point $x = \xi$;

(ii) $G(x, \xi)$ satisfies both boundary conditions (10);

(iii) although $G(x, \xi)$ is everywhere continuous for $0 \leqslant x \leqslant 1$, its derivative $G'_x(x, \xi)$ with respect to x is continuous only for $0 \leqslant x < \xi$ and $\xi < x \leqslant 1$, and has a jump discontinuity of $-1/p(\xi)$† at $x = \xi$, that is,

$$G'_x(\xi + 0, \xi) - G'_x(\xi - 0, \xi) = -\frac{1}{p(\xi)}. \quad (13)$$

If we can determine the fundamental set $y_1(x)$, $y_2(x)$ of equation (6),‡ then we can easily construct Green's function: Let

$$G(x, \xi) = \begin{cases} C_1(\xi)\,y_1(x) + C_2(\xi)\,y_2(x) & (0 \leqslant x < \xi), \\ D_1(\xi)\,y_1(x) + D_2(\xi)\,y_2(x) & (\xi < x \leqslant 1), \end{cases}$$

where $C_1(\xi)$, $C_2(\xi)$, $D_1(\xi)$ and $D_2(\xi)$ are four unknown functions of ξ, which can be determined from boundary conditions (10), the continuity condition $G(\xi + 0,\ \xi) = G(\xi - 0,\ \xi)$ and (13).

† Be careful. Some authors (including myself in [48]) consider this jump with opposite sign and this implies a change of G into $-G$.

‡ Thus it is necessary to integrate equation (6), not (the more general) equation (12).

These four equalities give us the algebraic non-homogeneous system

$$
\begin{cases}
C_1(\xi)\mathscr{A}_0[y_1(0)] + C_2(\xi)\mathscr{A}_0[y_2(0)] - D_1(\xi)\mathscr{B}_0[y_1(1)] \\
\qquad\qquad\qquad\qquad - D_2(\xi)\mathscr{B}_0[y_2(1)] = 0, \\
C_1(\xi)\mathscr{A}_1[y_1(0)] + C_2(\xi)\mathscr{A}_1[y_2(0)] - D_1(\xi)\mathscr{B}_1[y_1(1)] \\
\qquad\qquad\qquad\qquad - D_2(\xi)\mathscr{B}_1[y_2(1)] = 0, \\
-C_1(\xi)\,y_1(\xi) - C_2(\xi)\,y_2(\xi) + D_1(\xi)\,y_1(\xi) + D_2(\xi)\,y_2(\xi) = 0, \\
-C_1(\xi)\,y_1'(\xi) - C_2(\xi)\,y_2'(\xi) + D_1(\xi)\,y_1'(\xi) + D_2(\xi)\,y_2'(\xi) = -1/p(\xi)
\end{cases}
$$

whose determinant is equal to†

$$
\begin{vmatrix} y_1(\xi) & y_2(\xi) \\ y_1'(\xi) & y_2'(\xi) \end{vmatrix} \Delta,
$$

and consequently, by virtue of (11), does not vanish.

For instance, Green's function for the Sturm-Liouville system

$$
\frac{d^2 y}{dx^2} + \lambda r(x)\,y = 0, \quad y(0) = y(1) = 0
$$

is the previous *triangular kernel* (3·10-16).

Now let $z(x) = G(x, \xi)$ and integrate both sides of (9) over the basic interval (0, 1). Taking into account the discontinuity of G'_x, we get

$$
\int_0^1 \frac{d}{dx}[p(x)\,y(x)\,G'_x(x,\xi)]\,dx
$$
$$
= [p(x)\,y(x)\,G'_x(x,\xi)]_0^1 - p(\xi)\,y(\xi)\,[G'_x(\xi+0,\xi) - G'_x(\xi-0,\xi)]
$$
$$
= [p(x)\,y(x)\,G'_x(x,\xi)]_0^1 + y(\xi).
$$

Therefore, we obtain

$$
\int G(x,\xi)\mathscr{L}[y(x)]\,dx = \left[p(x)\left(G\frac{dy}{dx} - y\frac{\partial G}{\partial x}\right)\right]_0^1 - y(\xi). \tag{14}
$$

This formula can be simplified by observing that, by virtue of the rule for multiplication of determinants (in this case, multiplication of *rows times columns*), we have

$$
\begin{vmatrix} a_{00} & a_{01} \\ a_{10} & a_{11} \end{vmatrix}
\begin{vmatrix} y(0) & z(0) \\ y'(0) & z'(0) \end{vmatrix}
=
\begin{vmatrix} \mathscr{A}_0[y(0)] & \mathscr{A}_0[z(0)] \\ \mathscr{A}_1[y(0)] & \mathscr{A}_1[z(0)] \end{vmatrix},
$$

$$
\begin{vmatrix} b_{00} & b_{01} \\ b_{10} & b_{11} \end{vmatrix}
\begin{vmatrix} y(1) & z(1) \\ y'(1) & z'(1) \end{vmatrix}
=
\begin{vmatrix} \mathscr{B}_0[y(1)] & \mathscr{B}_0[z(1)] \\ \mathscr{B}_1[y(1)] & \mathscr{B}_1[z(1)] \end{vmatrix}.
$$

† To obtain this result, we need only add the third column to the first, and the fourth column to the second.

If both functions y and z satisfy boundary conditions (10), then the two determinants on the right are equal, and, setting

$$\begin{vmatrix} a_{00} & a_{01} \\ a_{10} & a_{11} \end{vmatrix} = A, \qquad \begin{vmatrix} b_{00} & b_{01} \\ b_{10} & b_{11} \end{vmatrix} = B, \tag{15}$$

we obtain the equality

$$A[y(0)\,z'(0) - y'(0)\,z(0)] = B[y(1)\,z'(1) - y'(1)\,z(1)]. \tag{16}$$

Consequently, if $z(x)$ is identified with $G(x, \xi)$, equation (14) assumes the simplified form

$$\int G(x, \xi)\,\mathscr{L}[y(x)]\,dx = -y(\xi), \tag{17}$$

provided (as often happens) the *key condition*

$$Ap(1) = Bp(0) \tag{18}$$

is satisfied.†

In this case, system (12) + (10) is reduced to a Fredholm integral equation by substituting for $\mathscr{L}[y(x)]$ in (17) the value $-\lambda r(x)\,y(x)$ given by (12). We obtain the homogeneous integral equation

$$y(\xi) - \lambda \int_0^1 G(x, \xi)\,r(x)\,y(x)\,dx = 0. \tag{19}$$

Because of the factor $r(x)$, the kernel of this equation is not symmetric. However, Green's function $G(x, \xi)$ *is symmetric.*

In fact, if we set

$$y(x) = G(x, \xi), \quad z(x) = G(x, \eta) \quad (\xi \neq \eta)$$

in Green's formula (9), an integration similar to that which gave us (14) gives

$$0 = [p(x)\,\{G(x, \eta)\,G'_x(x, \xi) - G(x, \xi)\,G'_x(x, \eta)\}]_0^1 + G(\xi, \eta) - G(\eta, \xi).$$

If the key condition (18) is satisfied, then the first term on the right is zero, since $G(x, \xi)$ and $G(x, \eta)$ both satisfy boundary condition (10). Hence we have

$$G(\xi, \eta) = G(\eta, \xi). \tag{20}$$

If the *function $r(x)$ never vanishes,* i.e. if

$$r(x) > 0 \tag{21}$$

---

† When the key condition (18) is satisfied, Ince [21], p. 217, and other authors speak of *self-adjoint boundary conditions.*

(if $r(x) < 0$, change $r(x)$ to $-r(x)$ and λ to $-\lambda$), then (19) can be reduced to an integral equation with a *symmetric kernel* by setting

$$y(x) \sqrt{[r(x)]} = \phi(x). \tag{22}$$

We get, if we return to the usual variables x, y, the integral equation

$$\phi(x) - \lambda \int K(x, y) \, \phi(y) \, dy = 0 \tag{23}$$

with the symmetric kernel

$$K(x, y) = \sqrt{[r(x) \, r(y)]} \, G(x, y). \tag{24}$$

From this and from considerations which follow, we obtain the following important theorem:

If $p(x) > 0$, if (11), (18) and (21) hold, and if the symmetric kernel (24) is quadratically integrable, then the Sturm-Liouville system (12) + (10) has non-trivial solutions only for a denumerable number of real values $\lambda_1, \lambda_2, \lambda_3, \ldots$ of λ: the eigenvalues of the symmetric kernel (24). To each eigenvalue there correspond at most two linearly independent 'eigensolutions' of the system, which are connected with the eigenfunctions of the kernel by equality (22).

All the eigenvalues of the kernel (24) are *at most* of index 2, because it is impossible to have more than *two* linearly independent solutions of the same *second* order linear differential equation. Moreover, we can sometimes assert that the eigenvalues are necessarily *simple*; for instance, if $\mathscr{A}_1 \equiv \mathscr{B}_0 \equiv 0$, i.e. if

$$a_{10} = a_{11} = b_{00} = b_{01} = 0 \tag{25}$$

(an important case in which the key condition (18) is automatically satisfied because $A = B = 0$). In fact, there can exist only a *simple* infinity of solutions of a linear differential equation of the second order satisfying a condition of the type

$$a_{00} y(0) + a_{01} y'(0) = 0.†$$

On the other hand, there are certainly *an infinite number of eigenvalues*, for otherwise there would be only a finite number of

---

† Because all the solutions proportional to the one defined by the initial values $y(0) = a$, $y'(0) = b$ ($a_{00} a + a_{01} b \neq 0$) do not satisfy the condition.

eigenfunctions $\phi_h(x)$, and their ON-system would be necessarily *open*, in contradiction to the following:

COMPLETENESS THEOREM. *The system of the eigenfunctions* $\{\phi_h(x)\}$ *of a Sturm-Liouville system of the type* $(10) + (12)$ *is complete in the entire L_2-space.*

For the proof we shall show that

(i) each D_2-*function* (i.e. each twice differentiable function $f(x)$ satisfying boundary conditions (10), whose second derivative $f''(x)$ belongs to the class L_2) can be uniformly and arbitrarily closely approximated by the functions $\phi_h(x)$;

(ii) each L_2-function can be arbitrarily closely approximated by D_2-functions.

To prove (i) it is sufficient to observe that, by virtue of the Hilbert-Schmidt theorem, each D_2-function can be expanded in an absolutely and uniformly convergent series of the form $\Sigma a_h \phi_h(x)$, because, if we put

$$g(x) = \mathscr{L}[f(x)] = \frac{d}{dx}\left(p\,\frac{df}{dx}\right) + qf,$$

then it follows from (17) that

$$f(x) = -\int G(x,y)\,g(y)\,dy,$$

that is, $$f(x) = -\frac{1}{\sqrt{[r(x)]}}\int K(x,y)\,\frac{g(y)}{\sqrt{[r(y)]}}\,dy.$$

To prove (ii) we first observe that every L_2-function can be arbitrarily closely approximated (in the mean) by means of suitable polynomials, because the system of powers is complete in any finite interval (§ 3·5). Furthermore, every polynomial can be arbitrarily closely approximated (in the mean) by suitable D_2-functions. To do this it is sufficient to modify slightly the polynomial (without destroying its twice differentiability) in two arbitrarily small neighborhoods of the end-points in order to satisfy boundary conditions (10).

The *completeness theorem* shows that the class of kernels arising from Sturm-Liouville systems is quite special, because, in general, nothing similar is true for an arbitrary symmetric L_2-kernel or for symmetric kernels for which Mercer's theorem is valid. Furthermore, while the only general 'asymptotic property' of the eigenvalues λ_n of a symmetric L_2-kernel is the convergence of the series

$\Sigma\lambda_n^{-2}$ (cf. §3·10), the eigenvalues arising from a Sturm-Liouville system and the corresponding eigenfunctions have very precise asymptotic properties. For instance, it can be proved that

$$\lambda_n = O(n^2) \quad \text{as} \quad n \to \infty.$$

Since this subject belongs properly to the theory of differential equations (even though the main device is the reduction of the given differential equation to a Volterra integral equation),† we shall limit ourselves here to reporting (without proof) some interesting results for the important case (25). We shall further assume that $p(x) \equiv r(x) \equiv 1$, i.e. that the differential equation has the simplified form

$$\frac{d^2y}{dx^2} + [q(x) + \lambda]y = 0. \tag{26}$$

(This can always be accomplished by means of elementary transformations.)

Putting

$$\int_0^1 q(x) = Q,$$

we have‡

$$\sqrt{\lambda_n} = \begin{cases} n\pi - \dfrac{\pi}{n}\left(\dfrac{a_{00}}{a_{01}} - \dfrac{b_{10}}{b_{11}} + \dfrac{1}{2}Q\right) + O(n^{-2}) & (a_{01}b_{11} \neq 0), \\[2ex] (n+\tfrac{1}{2})\pi - \dfrac{\pi}{n}\left(-\dfrac{b_{10}}{b_{11}} + \dfrac{1}{2}Q\right) + O(n^{-2}) & (a_{01}=0,\ b_{11} \neq 0), \\[2ex] (n+\tfrac{1}{2})\pi - \dfrac{\pi}{n}\left(\dfrac{a_{00}}{a_{01}} + \dfrac{1}{2}Q\right) + O(n^{-2}) & (a_{01} \neq 0,\ b_{11}=0), \\[2ex] (n+1)\pi - \dfrac{\pi}{2n}Q + O(n^{-2}) & (a_{01}=b_{11}=0). \end{cases} \tag{27}$$

For the corresponding eigenfunctions, we have the formulae§

$$\phi_n(x) = \begin{cases} \sqrt{2}\cos(n\pi x) + O(n^{-1}) & (a_{01}b_{11} \neq 0), \\ \sqrt{2}\cos(n+\tfrac{1}{2})\pi x + O(n^{-1}) & (a_{01}=0,\ b_{11} \neq 0), \\ \sqrt{2}\sin(n+\tfrac{1}{2})\pi x + O(n^{-1}) & (a_{01} \neq 0,\ b_{11}=0), \\ \sqrt{2}\sin(n+1)\pi x + O(n^{-1}) & (a_{01}=b_{11}=0). \end{cases} \tag{28}$$

† See, for example, Tricomi [46], p. 175 or Ince [21], p. 270.

‡ Remember that the numbers a_{ih}, b_{ih} are the coefficients appearing in boundary conditions (10).

§ In my book [46], the second terms (of order n^{-1}) of the formulae (28) are explicitly given as well as the asymptotic expressions of the first derivative $\phi_n'(x)$ of $\phi_n(x)$.

Finally, we note that by using Green's function $G(x, \xi)$ we can put the solution of the *non-homogeneous* differential equation

$$\frac{d}{dx}\left[p(x)\frac{dy}{dx}\right] + q(x)\, y = f(x) \qquad (29)$$

satisfying boundary conditions (10) into the elegant form

$$y(x) = -\int G(x, \xi) f(\xi)\, d\xi. \qquad (30)$$

This is an immediate consequence of (17) since equation (29) assures us that $\mathscr{L}[y(x)] = f(x)$.

3·14. Critical Velocities of a Rotating Shaft and Transverse Oscillations of a Beam

In the first chapter (§§ 1·1 and 1·10) we saw that the two mechanical problems mentioned in the title can be reduced to *the same* homogeneous Fredholm integral equation of the second kind, i.e. to the equation

$$z(x) - \omega^2 \int_0^1 G(x, y)\, \mu(y)\, z(y)\, dy = 0. \qquad (1)$$

Here ω is the angular velocity of the shaft (or, respectively, the angular frequency of the vibrations of the beam), $\mu(x)$ is the *linear density*, i.e. $\mu(x)\, dx$ is the mass between the cross-section at x and at $x + dx$, and $G(x, y)$ is the *influence function of the bar*.† Since $\mu(x) > 0$, if we put

$$\sqrt{[\mu(x)]}\, z(x) = \phi(x), \qquad (2)$$

we obtain a homogeneous integral equation

$$\phi(x) - \omega^2 \int K(x, y)\, \phi(y)\, dy = 0 \qquad (3)$$

with the symmetric kernel

$$K(x, y) = \sqrt{[\mu(x)\, \mu(y)]}\, G(x, y). \qquad (4)$$

In view of the previous section, this fact is sufficient to prove the existence of an infinite number of critical velocities (respectively, of

† Remember that $G(x, y)$ gives the displacement (parallel to the z-axis) of the cross-section at x if a concentrated unit load acts (in the direction of the z-axis) at the point $x = y$. The coincidence of the letter G with the symbol for Green's function is not accidental, since the influence function can be considered as Green's function for the differential equation of the fourth order (which we shall not consider here) governing the dynamics of a bar.

natural frequencies) $\omega_h = \sqrt{\lambda_h}$ related to the (positive) eigenvalues λ_h of the symmetric kernel (4).

Furthermore, we can now also obtain upper and lower bounds for these eigenvalues, etc. Before going into detail, we shall emphasize some general properties of the *kernel of elasticity theory*, i.e. of the influence function $G(x, y)$ of a beam considered as a kernel of a Fredholm integral equation.†

A first, quite obvious but interesting, property is that *Fredholm's operator of the first kind*

$$\mathscr{F}_x^{(1)}[\phi(y)] \equiv \int_0^1 K(x, y)\, \phi(y)\, dy,$$

in the case $K \equiv G$, can be considered as a functional operator which transforms a system of distributed loads $\phi(x)$ acting (parallel to the z-axis) on the beam, into the corresponding *elastica* of it, in the sense that, after deformation, the shape of the axis of the beam (which coincides initially with the segment $0 \leqslant x \leqslant 1$) will be the curve of the xz-plane represented by the equation

$$z = \mathscr{F}_x^{(1)}[\phi(y)].$$

A consequence is that, in this field, the Hilbert-Schmidt theorem can be stated expressively as follows: *Any function which can be represented graphically by the elastica of a certain beam bent under the action of suitable (quadratically integrable) loads, can be expanded in an absolutely and uniformly convergent series in terms of the eigenfunctions of the influence function of the beam.*

The *positive character* of the kernel $G(x, y)$ in the sense of § 3·12 is also of interest.

In fact, Hilbert's quadratic form

$$J(\phi, \phi) = \iint G(x, y)\, \phi(x)\, \phi(y)\, dx\, dy = \int \mathscr{F}_x^{(1)}[\phi(y)]\, \phi(x)\, dx$$

coincides with the *work* of the elastic forces under the deformation produced by the loads $\phi(x)$; consequently, its values are 'always' positive.

Since the λ_h's are all positive, there is no doubt that to each eigenvalue λ_h of the previous integral equation (3) corresponds an *effective* critical angular velocity (respectively, a natural angular frequency) $\omega_h = \sqrt{\lambda_h}$.

† In view of the character of these properties, the passage from the kernel $G(x, y)$ to the more general one of type (4) is immediate.

We now determine the influence function $G(x, y)$, starting with the case of a bar *simply supported* at both ends $x = 0$ and $x = 1$.

If R_0 and R_1 are, respectively, the (*a priori* unknown) reactions at the supports under the action of a unit load at $x = y$ (see Fig. 9), the *bending moment* M at a generic point x of the bar is obviously given by the formulae

$$M = \begin{cases} -R_0 x & (0 \leqslant x \leqslant y), \\ -R_1(1-x) & (y \leqslant x \leqslant 1). \end{cases}$$

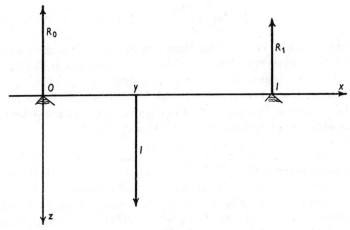

Fig. 9

For the equilibrium of the system of the three forces R_0, R_1, 1 we must have

$$1 - R_0 - R_1 = 0, \quad R_1 . 1 = 1 . y.$$

Hence, $R_1 = y$, $R_0 = 1 - y$, and we see that the bending moment coincides with the triangular kernel T of § 3·10 (with a change of sign), i.e. that

$$M = -T(x, y). \tag{5}$$

Consequently, the influence function $G(x, y)$ must satisfy the differential equation†

$$E_x I_x \frac{\partial^2 G}{\partial x^2} = -T(x, y), \tag{6}$$

† This is the so-called *equation for the bending moment*; see, for example, J. Prescott, *Applied Elasticity*, Longmans, Green and Co., London, etc., 1924, p. 51.

together with the boundary conditions

$$G(0, y) \equiv G(1, y) \equiv 0, \tag{7}$$

where E_x, I_x denote, respectively, Young's modulus and the moment of inertia of the cross-section of the bar (with respect to the *neutral axis* perpendicular to the xz-plane) at x.

The system $(6) + (7)$ can be solved elegantly using the definite integral†

$$T^* = \int_0^1 T(x, z) \, T(z, y) \, F(z) \, dz, \tag{8}$$

where

$$F(x) = \frac{1}{E_x I_x}. \tag{9}$$

In fact, T^* obviously satisfies boundary conditions (7) since $T(0, y) \equiv T(1, y) \equiv 0$. It also satisfies differential equation (6) because T^* is a *symmetric* function of x and y which, for $x \leqslant y$, has the following explicit expression:

$$T^* = xy \int_0^1 (1-z)^2 F(z) \, dz - x \int_0^y (1-y)(y-z) F(z) \, dz$$
$$- (1-y) \int_0^x z(x-z) F(z) \, dz,$$

from which it follows that

$$\frac{\partial T^*}{\partial x} = y \int_0^1 (1-z)^2 F(z) \, dz - \int_0^y (1-y)(y-z) F(z) \, dz$$
$$- (1-y) \int_0^x z F(z) \, dz$$

and

$$\frac{\partial^2 T^*}{\partial x^2} = -x(1-y) F(x) = -\frac{T(x, y)}{E_x I_x} \quad (x \leqslant y).$$

Hence, for a beam simply supported at the ends, we have

$$G(x, y) = \int_0^1 T(x, z) \, T(z, y) \, F(z) \, dz. \tag{10}$$

Similarly, for a beam *clamped* at $x = 0$ and *free* at $x = 1$, we have

$$M = y - x \quad (0 \leqslant x \leqslant y), \qquad M = 0 \quad (y \leqslant x \leqslant 1);$$

† This form of the solution is suggested by a general property of Green's function for some differential equations of higher order. This is indicated in my paper: 'Sulla funzione di Green di un'equazione differenziale decomposta in fattori simbolici', *Atti R. Accad. Sci. Torino*, 80, 1944–5, pp. 159–83.

consequently, we obtain

$$\frac{\partial^2 G}{dx^2} = (y-x)\,F(x), \quad G(0,y) \equiv G'_x(0,y) \equiv 0 \quad (0 \leqslant x \leqslant y),$$

and successively

$$G(x,y) = \int_0^x (x-z)\,(y-z)\,F(z)\,dz \quad (0 \leqslant x \leqslant y).$$

Even better, since $G(x,y) = G(y,x)$, we can now write

$$G(x,y) = \int_0^{\min(x,y)} (x-z)\,(y-z)\,F(z)\,dz. \tag{11}$$

In conclusion, if we introduce the dimensionless functions

$$m(x) = \frac{\mu(x)}{\mu(0)}, \quad f(x) = \frac{F(x)}{F(0)} = \frac{E_0 I_0}{E_x I_x} \tag{12}$$

and put

$$\omega^2 \mu(0)\,F(0) = \lambda, \tag{13}$$

we can state that *the critical angular velocities* ω_h *of a rotating shaft (or, respectively, the natural angular frequencies* ω_h *of a vibrating beam) are given by the formulae*

$$\omega_h = \sqrt{\left(\frac{\lambda_h}{\mu(0)\,F(0)}\right)} = \sqrt{\left(\frac{E_0 I_0}{\mu(0)}\lambda_h\right)} \quad (h = 1, 2, 3, \dots), \tag{14}$$

where $\lambda_1, \lambda_2, \dots$ *denote the successive eigenvalues of the positive Fredholm kernel*

$$K'(x,y) = \sqrt{[m(x)\,m(y)]} \int_0^1 T(x,z)\,T(z,y)\,f(z)\,dz \tag{15}$$

if the beam is simply supported, or of the kernel

$$K''(x,y) = \sqrt{[m(x)\,m(y)]} \int_0^{\min(x,y)} (x-z)\,(y-z)\,f(z)\,dz \tag{16}$$

if one end $(x = 0)$ *is clamped and the other end* $(x = 1)$ *is free.*

For instance, for a *simply supported homogeneous beam,* i.e. for the case involving the kernel (15) with $m(x) \equiv f(x) \equiv 1$, by virtue of (3·11-17), we have

$$K'(x,y) = \int_0^1 T(x,z)\,T(z,y)\,dz = T_2(x,y)$$

$$= \frac{2}{\pi^4} \sum_{h=1}^{\infty} h^{-4} \sin(h\pi x)\sin(h\pi y),$$

and thus we obtain

$$\lambda_h = (\pi h)^4 \quad (h = 1, 2, 3, \ldots) \tag{17}$$

and consequently

$$\omega_h = \sqrt{\left(\frac{EI}{\mu}\right)} \pi^2 h^2 \quad (h = 1, 2, 3, \ldots). \tag{18}$$

However, in the general case, and even in the case (§ 1·10) of a homogeneous ($m \equiv f \equiv 1$) but clamped beam, it is in general not possible to obtain explicit expressions for the eigenvalues λ_h. One must be content with approximate evaluations like those of § 3·11, which are usually sufficiently accurate for practical purposes.

For example, in the case of the *homogeneous clamped beam* we have

$$A_1 = \int_0^1 K''(x, x)\, dx = \frac{1}{3} \int_0^1 x^3\, dx = \frac{1}{12},$$

$$A_2 = \int_0^1 K_2''(x, x)\, dx = \int_0^1 \left(\frac{x^4}{12} - \frac{x^5}{12} + \frac{x^6}{36} - \frac{x^7}{630}\right) dx = \frac{11}{1680},$$

and consequently (3·11-16) gives us two different lower bounds

$$\lambda_1 > 12, \quad \lambda_1 > \sqrt{\frac{1680}{11}} = 12 \cdot 358$$

for the first eigenvalue $\lambda_1 = 12 \cdot 362$ (cf. § 1·10). The second bound is very good.

Several years ago using similar methods, I studied† the vibrations of a simply supported beam having the shape of a truncated cone (briefly TC) or the shape of a wedge (briefly W).

In these technically important cases we have, respectively,

$$m(x) = (1 - \theta x)^2, \quad f(x) = (1 - \theta x)^{-4} \quad \text{(TC)}$$

and

$$m(x) = 1 - \theta x, \quad f(x) = (1 - \theta x)^{-3} \quad \text{(W)},$$

where θ is a given positive constant less than 1. In both cases the differential equation of the fourth order governing the dynamics of the beam can be integrated explicitly by means of Bessel functions.‡ In spite of this fact, it does not seem easy to obtain simple expressions for the eigenvalues λ_h in this way. However, by improving

† F. Tricomi, 'Sulle vibrazioni trasversali di aste, specialmente di bielle, di sezione variabile', *Ricerche di Ingegneria*, Roma, **4**, 1936, pp. 47–53.

‡ Kirchhoff, *Gesammelte Abhandlungen*, Barth, Leipzig, 1882, p. 339.

formula (3·11-16) for $m = 1$ (for details, see my quoted paper), we obtain the simple formula

$$\frac{1}{\lambda_1} \leqslant \frac{1}{\lambda_1'} = \mathscr{T} - \left(\frac{1}{90} - \frac{1}{\pi^4}\right)(1-\theta)^p, \tag{19}$$

where $p = 2$ for (TC) and $p = 1$ for (W), and

$$\mathscr{T} = \int_0^1 \int_0^1 T^2(x, y)\, m(x) f(y)\, dx\, dy \tag{20}$$

is equal to $[90(1-\theta)]^{-1}$ in the case (TC) and is equal to

$$-[1/(36\theta^5)]\,[12\theta(1-\theta) + 7\theta^3 + 3(2-\theta)(2-\theta+\theta^2)\log(1-\theta)]$$

in the case (W).

On the other hand, with devices similar to those used in § 3·11 to obtain formulae (3·11-14) and (3·11-15), we easily obtain the upper bound

$$\lambda_1 \leqslant \lambda_1'' = \begin{cases} \pi^4 \dfrac{1 - 2\theta + 6H_2\theta^2 - 4H_3\theta^3 + H_4\theta^4}{1 - \theta + H_2\theta^2} & \text{(TC)}, \\[2ex] \pi^4 \dfrac{2 - 3\theta + 6H_2\theta^2 - 2H_3\theta^3}{2 - \theta} & \text{(W)}, \end{cases} \tag{21}$$

where

$$H_n = 2 \int_0^1 x^n \sin^2 \pi x\, dx \quad (n = 1, 2, 3, 4). \tag{22}$$

These formulae give good numerical results. This is shown by the following table extracted from my paper mentioned above:

$\theta \to$		0·0	0·1	0·2	0·3	0·4	0·5	0·6	0·7
TC	$\sqrt{\lambda_1'}/\pi^2$	1·0000	0·9383	0·8770	0·8150	0·7523	0·6837	0·6094	0·5269
	$\sqrt{\lambda_1''}/\pi^2$	1·0000	0·9509	0·9037	0·8586	0·8161	0·7765	0·7403	0·7079
W	$\sqrt{\lambda_1'}/\pi^2$	1·0000	0·9419	0·8836	0·8250	0·7654	0·7014	0·6353	0·5636
	$\sqrt{\lambda_1''}/\pi^2$	1·0000	0·9505	0·9022	0·8552	0·8097	0·7660	0·7248	0·6860

The case of a *connecting rod* having the shape of a truncated cone with $\theta = 0·19$ was studied by my late colleague P. E. Brunelli,[†] who, after *three months of calculation* with an ingenious but extremely tedious 'elementary' method, found

$$\sqrt{\lambda_1} = \pi^2\, 0·8977. \tag{23}$$

† 'Oscillazioni di bielle di sezione variabile', *Atti R. Ist. d'Incoraggiamento di Napoli*, 1929.

With the previous formulae we easily obtain

$$\sqrt{\lambda_1'/\pi^2} = 0\cdot883, \quad \sqrt{\lambda_1''/\pi^2} = 0\cdot908;$$

hence, in the mean

$$\sqrt{\lambda_1} = \pi^2 (0\cdot896 \pm 0\cdot013). \tag{24}$$

With modern computers the three months of calculation could surely be shortened, but I believe that the method which furnished the value (24) is still the shorter.

3·15. Symmetric Fredholm Equations of the First Kind

Some mathematicians still have a kind of fear whenever they encounter a Fredholm integral equation of the *first* kind:

$$\int_0^1 K(x,y)\,\phi(y)\,dy = f(x) \quad (0 \leqslant x \leqslant 1). \tag{1}$$

Today this fear is no longer justified, especially for the *symmetric case* $K(x,y) = K(y,x)$. Here the Hilbert-Schmidt theorem gives a simple way of discussing equation (1) thoroughly in L_2-space, provided we can determine the eigenvalues λ_h and eigenfunctions $\phi_h(x)$ of the kernel $K(x,y)$, i.e. provided we can master the corresponding equation of the *second* kind:

$$\phi(x) - \lambda \int K(x,y)\,\phi(y)\,dy = f(x).$$

To be precise, the considerations of § 3·10 (Hilbert-Schmidt theorem) show that if equation (1) has a solution $\phi(x)$ of class L_2, then

$$a_h = \frac{\xi_h}{\lambda_h} \quad (h = 1, 2, 3, \ldots),$$

where

$$\xi_h = \int \phi(x)\,\phi_h(x)\,dx$$

are the unknown Fourier coefficients of $\phi(x)$ with respect to the system $\{\phi_h(x)\}$,

$$a_h = \int f(x)\,\phi_h(x)\,dx,$$

are the corresponding coefficients of the given function $f(x)$, and the λ_h's are the eigenvalues of the kernel. From this it follows that

$$\xi_h = a_h \lambda_h. \tag{2}$$

Hence, by virtue of the Riesz-Fischer theorem, there are only two possibilities: either

(i) the infinite series
$$\sum_{h=1}^{\infty} a_h^2 \lambda_h^2 \tag{3}$$

diverges, and our equation has no solution of class L_2; or

(ii) series (3) converges and there is at least one L_2-function $\phi_0(x)$ which satisfies our equation, and it can be calculated as the limit in the mean

$$\phi_0(x) = \text{l.i.m.}_{n\to\infty} \sum_{h=1}^{n} a_h \lambda_h \phi_h(x). \tag{4}$$

If the system of eigenfunctions $\{\phi_h(x)\}$ is closed, i.e. if the kernel $K(x, y)$ is closed, then the function $\phi_0(x)$ is the *unique* solution of (1) (neglecting functions which vanish almost everywhere). However, if $\Phi_1(x), \Phi_2(x), \ldots$ are non-null functions orthogonal to all the functions $\{\phi_h\}$, then the function $\phi_0(x) + C_1 \Phi_1(x) + C_2 \Phi_2(x) + \ldots$, where the C_i are arbitrary constants, is also a solution of (1) and vice versa.

For example, the equation of the first kind

$$\int_0^1 T(x, y)\, \phi(y)\, dy = f(x), \tag{5}$$

where $T(x, y)$ is the earlier triangular kernel, for which (§ 3·10)

$$\lambda_h = (h\pi)^2, \quad \phi_h(x) = \sqrt{2} \sin(h\pi x) \quad (h = 1, 2, 3, \ldots),$$

has one (and only one) solution of class L_2 if and only if, when we set

$$a_h = \sqrt{2} \int_0^1 f(x) \sin(h\pi x)\, dx,$$

the infinite series
$$\sum_{h=1}^{\infty} a_h^2 \lambda_h^2 = \pi^2 \sum_{h=1}^{\infty} (h^2 a_h)^2 \tag{6}$$

converges. A strong condition indeed!

Similarly, *Poisson's transformation*

$$f(\theta) = \frac{1-\rho^2}{2\pi} \int_0^{2\pi} \frac{\phi(\alpha)\, d\alpha}{1 - 2\rho \cos(\theta - \alpha) + \rho^2} \quad (0 \leqslant \theta \leqslant 2\pi,\ 0 < \rho < 1) \tag{7}$$

(i.e. Poisson's integral for a fixed ρ) can be *inverted* in L_2-space if and only if the infinite series

$$\sum_{n=1}^{\infty} \frac{a_n^2 + b_n^2}{\rho^{2n}}, \tag{8}$$

where $\quad a_n = \frac{1}{\pi} \int_0^{2\pi} f(\theta) \cos n\theta \, d\theta, \quad b_n = \frac{1}{\pi} \int_0^{2\pi} f(\theta) \sin n\theta \, d\theta,$ $\qquad$ (9)

converges. In fact, the symmetric kernel

$$K(\theta, \alpha) = \frac{1-\rho^2}{2\pi} [1 - 2\rho \cos(\theta - \alpha) + \rho^2]^{-1}$$

$$= \frac{1}{2\pi} + \frac{1}{\pi} \sum_{h=1}^{\infty} \rho^h \cos h(\theta - \alpha), \qquad (10)$$

has, in the basic interval $(0, 2\pi)$, the eigenvalues

$$\lambda_0 = 1, \quad \lambda_{2h-1} = \lambda_{2h} = \rho^{-h} \quad (h = 1, 2, 3, \ldots), \qquad (11)$$

with corresponding eigenfunctions

$$\phi_0(x) = \frac{1}{\sqrt{(2\pi)}}, \quad \phi_{2h-1}(x) = \frac{1}{\sqrt{\pi}} \cos hx, \quad \phi_{2h}(x) = \frac{1}{\sqrt{\pi}} \sin hx$$
$$(h = 1, 2, 3, \ldots). \qquad (12)$$

Since infinite series (10) is absolutely and uniformly convergent for $|\rho| < 1$, we have

$$\int_0^{2\pi} K(\theta, \alpha) \, d\alpha = 1, \quad \int_0^{2\pi} K(\theta, \alpha) \frac{\cos}{\sin} n\alpha \, d\rho = \rho^n \frac{\cos}{\sin} n\theta \quad (n = 1, 2, 3, \ldots).$$

The corresponding case of Poisson's integral for the half-plane, i.e. of the integral equation

$$f(x) = \frac{\eta}{\pi} \int_{-\infty}^{\infty} \frac{\phi(\xi) \, d\xi}{(x - \xi)^2 + \eta^2} \quad (\eta > 0),$$

has been studied by H. Bateman.†

3·16. Reduction of a Fredholm Equation to a Similar One with a Symmetric Kernel

The remarkable properties of Fredholm integral equations with *symmetric* kernels show the desirability of reducing, if possible, a given integral equation to one with a symmetric kernel. Sometimes

† 'Some integral equations of potential theory', *J. Appl. Phys.* **17**, 1946, pp. 91–102.

this reduction is feasible with simple devices, as, for example, we have seen in the last two sections.

More generally, if the integral equation

$$\phi(x) - \lambda \int K(x,y)\,\phi(y)\,dy = f(x) \tag{1}$$

is given, multiplying by $K(x,z)$ and integrating with respect to x, we obtain

$$\int K(x,z)\,\phi(x)\,dx - \lambda \int K(x,z)\,dx \int K(x,y)\,\phi(y)\,dy = \int K(x,z)f(x)\,dx.$$

With a trivial change of notation, this can be written

$$\int [K(y,x) - \lambda K_L(x,y)]\,\phi(y)\,dy = \int K(y,x)f(y)\,dy, \tag{2}$$

where

$$K_L(x,y) = \int K(z,x)\,K(z,y)\,dz \tag{3}$$

is the so-called *left iterated* kernel of $K(x,y)$ and is obviously a symmetric kernel. Multiplying (2) by $-\lambda$ and then adding (1), we obtain the new Fredholm integral equation

$$\phi(x) - \lambda \int [K(x,y) + K(y,x) - \lambda K_L(x,y)]\,\phi(y)\,dy$$
$$= f(x) - \lambda \int K(y,x)f(y)\,dy, \tag{4}$$

whose kernel is symmetric, but not independent of λ.†

Similarly, starting with the associated equation

$$\psi(x) - \lambda \int K(y,x)\,\psi(y)\,dy = g(x), \tag{5}$$

we obtain another equation with a symmetric kernel

$$\psi(x) - \lambda \int [K(x,y) + K(y,x) - \lambda K_R(x,y)]\,\psi(y)\,dy$$
$$= g(x) - \lambda \int K(x,y)\,g(y)\,dy, \tag{6}$$

† The case in which the kernel depends upon λ, and especially when this dependence is linear, was studied by J. D. Tamarkin, *Ann. Math.* (2), **28**, 1927, pp. 127–52; C. Miranda, *Rend. Circ. Mat. Palermo*, **60**, 1936, pp. 286–304; *Mem. Accad. Sci. Torino*, (2), **70**, 1940, pp. 25–31; *Rend. Accad. Italia*, **2**, 1940; *Rend. Semin. Mat. Torino*, **12**, 1952–3, pp. 67–82; D. Greco, *Giorn. Mat. Battaglini*, **78**, 1948–9, pp. 216–37; **79**, 1949–50, pp. 86–120; R. Iglisch, *Math. Ann.* **117**, 1939, pp. 129–39, and others.

where K_R is the *right iterated* kernel

$$K_R(x,y) = \int K(x,z)\,K(y,z)\,dz. \tag{7}$$

The importance of the previous transformations lies not so much in equations (4) and (6), but rather in the introduction of the two symmetric kernels K_L and K_R associated with K. Consideration of these kernels permits us (as E. Schmidt noted)† to extend many results of the theory of integral equations with symmetric kernels to general equations. Moreover, we have already met the kernel

$$K(x,y) + K(y,x) - \lambda K_R(x,y) \tag{8}$$

in § 3·7 (Enskog's method).

We notice first that *both kernels K_L and K_R are positive kernels* since

$$\iint K_L(x,y)\,\phi(x)\,\phi(y)\,dx\,dy = \iiint K(z,x)\,K(z,y)\,\phi(x)\,\phi(y)\,dx\,dy\,dz$$

$$= \int dz \iint K(z,x)\,K(z,y)\,\phi(x)\,\phi(y)\,dx\,dy$$

$$= \int \left[\int K(z,x)\,\phi(x)\,dx \right]^2 dz \geqslant 0,$$

and similarly for $K_R(x,y)$.

Furthermore, we can easily prove that the kernels K_L and K_R have the same (positive) eigenvalues with the same indices. To be precise, if we call $\nu(x)$ any eigenfunction of $K_L(x,y)$ corresponding to the eigenvalue λ^2, and if we put

$$\mu(x) = \lambda \int K(x,y)\,\nu(y)\,dy, \tag{9}$$

then this new function $\mu(x)$ will be an eigenfunction of $K_R(x,y)$ corresponding to the same eigenvalue λ^2. Conversely, if $\mu(x)$ is an eigenfunction of $K_R(x,y)$ corresponding to the eigenvalue λ^2, then the function

$$\nu(x) = \lambda \int K(y,x)\,\mu(y)\,dy \tag{10}$$

is an eigenfunction of $K_L(x,y)$ for the same eigenvalue λ^2.

† E. Schmidt, *Math. Annalen*, **63**, 1907, pp. 433–76; **64**, 1907, pp. 161–74; and **65**, 1908, pp. 370–99.

In fact, from the equality

$$\nu(x) = \lambda^2 \int K_L(x,y)\,\nu(y)\,dy = \lambda^2 \int K(z,x)\,dz \int K(z,y)\,\nu(y)\,dy$$

with the use of (9), it follows that

$$\nu(x) = \lambda \int K(z,x)\,\mu(z)\,dz$$

and successively, by substitution of this expression into the right-hand side of equation (9), that

$$\mu(x) = \lambda^2 \int K(x,y)\,dy \int K(z,y)\,\mu(z)\,dz$$

$$= \lambda^2 \int \mu(z)\,dz \int K(x,y)\,K(z,y)\,dy$$

$$= \lambda^2 \int K_R(x,z)\,\mu(z)\,dz,$$

and vice versa.

Now we shall suppose that the common *spectrum* of the kernels K_L and K_R is

$$\lambda_1^2, \quad \lambda_2^2, \quad \lambda_3^2, \quad \ldots \tag{11}$$

with $$0 < \lambda_1 \leqslant \lambda_2 \leqslant \lambda_3 \leqslant \ldots,$$

and that the corresponding (orthonormalized) eigenfunctions of K_R are

$$\mu_1(x), \quad \mu_2(x), \quad \mu_3(x), \quad \ldots, \tag{12}$$

while $$\nu_1(x), \quad \nu_2(x), \quad \nu_3(x), \quad \ldots \tag{13}$$

are the corresponding eigenfunctions of K_L calculated by means of (10). Automatically they form an orthonormal system since

$$\int \nu_h(x)\,\nu_k(x)\,dx = \lambda_h \lambda_k \iiint K(y,x)\,K(z,x)\,\mu_h(y)\,\mu_k(z)\,dx\,dy\,dz$$

$$= \lambda_h \lambda_k \iint K_R(y,z)\,\mu_h(y)\,\mu_k(z)\,dy\,dz$$

$$= \frac{\lambda_h}{\lambda_k} \int \mu_h(y)\,\mu_k(y)\,dy = \begin{cases} 0 & (h \neq k), \\ 1 & (h = k). \end{cases}$$

Since K_R and K_L are positive, if we *assume that they are also continuous*, in view of Mercer's theorem we can write

$$K_R(x,y) = \sum_{h=1}^{\infty} \lambda_h^{-2} \mu_h(x) \mu_h(y), \quad K_L(x,y) = \sum_{h=1}^{\infty} \lambda_h^{-2} \nu_h(x) \nu_h(y),$$

(14)

and the absolute and uniform convergence of both series is assured. However, for the given kernel $K(x,y)$, we can only write

$$K(x,y) \sim \sum_{h=1}^{\infty} \frac{\mu_h(x) \nu_h(y)}{\lambda_h},$$

(15)

where Hurwitz's sign $\sim$ denotes (as usual) that the right-hand side is the *Fourier series* of $K(x,y)$, considered as a function of x or of y, with respect to the ON-system (12) or (13), respectively. But, as in § 3·9 (or even more easily than in § 3·9), we can show that the series on the right-hand side of (15) converges at least *in the mean* to $K(x,y)$, i.e. that

$$K(x,y) = \underset{n \to \infty}{\text{l.i.m.}} \sum_{h=1}^{n} \lambda_h^{-1} \mu_h(x) \nu_h(y).$$

(16)

In fact, from (3·2-7) and the second expansion of (14), we have

$$\int \left[K(x,y) - \sum_{h=1}^{n} \lambda_h^{-1} \mu_h(x) \nu_h(y) \right]^2 dx = \int K^2(x,y)\, dx - \sum_{h=1}^{n} \frac{\nu_h^2(y)}{\lambda_h^2}$$

$$= K_L(y,y) - \sum_{h=1}^{n} \frac{\nu_h^2(y)}{\lambda_h^2}$$

$$= \sum_{h=n+1}^{\infty} \frac{\nu_h^2(y)}{\lambda_h^2}.$$

Since series (14) converges uniformly, to any positive ϵ there corresponds a positive integer n_0, independent of y, such that for $n > n_0$ we have

$$\int \left[K(x,y) - \sum_{h=1}^{n} \lambda_h^{-1} \mu_h(x) \nu_h(y) \right]^2 dx < \epsilon,$$

in accord with (16).

Thus we see that Hurwitz's sign $\sim$ in (15) can be changed into an equality sign if the series on the right-hand side converges uniformly, but neither from (15) nor from the corresponding equality can it be deduced that the λ_h are eigenvalues of $K(x,y)$! This may even be a kernel without eigenvalues, like those of Volterra integral equations.

The main application of the ON-systems (12) and (13) is, however, the extension of the Hilbert-Schmidt theorem to non-symmetric kernels. Using the method of § 3·10, we can prove the following theorem:

If the kernel $K(x, y)$, considered as a function of y, is quadratically integrable, then any function $f(x)$ which can be represented by an integral of the form

$$f(x) = \int K(x, y) g(y) \, dy \tag{17}$$

where g is an arbitrary L_2-function, can also be represented by an absolutely and almost uniformly convergent series in the functions (12), *that is,*

$$f(x) = \sum_{h=1}^{\infty} a_h \mu_h(x) \quad with \quad a_h = \int f(x) \mu_h(x) \, dx = \frac{1}{\lambda_h} \int g(x) \nu_h(x) \, dx. \tag{18}$$

Similarly, if the kernel is a quadratically integrable function of x, then any function $f(x)$ which can be represented by an integral of the form

$$f(x) = \int K(y, x) g(y) \, dy, \tag{19}$$

can also be represented by an absolutely and almost uniformly convergent series of the form

$$f(x) = \sum_{h=1}^{\infty} b_h \nu_h(x) \quad with \quad b_h = \int f(x) \nu_h(x) \, dx = \frac{1}{\lambda_h} \int g(x) \mu_h(x) \, dx. \tag{20}$$

If the kernel belongs not only to the class L_2 but also to L_2^*, then the convergence of the above series is not only *almost uniform* but *uniform* as well.

This beautiful theorem allows us to generalize immediately the ideas of the previous section to any Fredholm integral equation of the *first kind* with an L_2-kernel. It is sufficient to replace the eigenfunctions $\{\phi_h(x)\}$ of the symmetric kernel by the functions $\{\mu_h(x)\}$.

3·17. Some Generalizations

The theory of Fredholm integral equations, which we have presented, permits many generalizations which leave its main features unchanged.

The theory may first be generalized by considering a *system* of Fredholm integral equations. Such a system can be easily reduced to a single equation. For instance, a system of two Fredholm integral equations

$$\phi_1(x) - \lambda \int_0^1 K_{1,1}(x,y)\,\phi_1(y)\,dy - \lambda \int_0^1 K_{1,2}(x,y)\,\phi_2(y)\,dy = f_1(x),$$

$$\phi_2(x) - \lambda \int_0^1 K_{2,1}(x,y)\,\phi_1(y)\,dy - \lambda \int_0^1 K_{2,2}(x,y)\,\phi_2(y)\,dy = f_2(x),$$

$$\tag{1}$$

in the basic interval $(0, 1)$ can be reduced to the following single equation in the basic interval $(0, 2)$:

$$\Phi(x) - \lambda \int_0^2 K(x,y)\,\Phi(y)\,dy = F(x), \tag{2}$$

where

$$\Phi(x) = \begin{cases} \phi_1(x) & (0 \leqslant x \leqslant 1), \\ \phi_2(x-1) & (1 < x \leqslant 2), \end{cases} \qquad F(x) = \begin{cases} f_1(x) & (0 \leqslant x < 1), \\ f_2(x-1) & (1 < x \leqslant 2), \end{cases}$$

and $K(x,y)$ is equal to

$$K_{1,1}(x,y), \quad K_{1,2}(x,y-1), \quad K_{2,1}(x-1,y), \quad K_{2,2}(x-1,y-1)$$

in the four squares

$$(0 \leqslant x < 1,\ 0 \leqslant y < 1), \quad (0 \leqslant x < 1,\ 1 < y \leqslant 2),$$
$$(1 < x \leqslant 2,\ 0 \leqslant y < 1), \quad (1 < x \leqslant 2,\ 1 < y \leqslant 2),$$

respectively.

Another observation is that, although up to now we have studied integral equations with a *finite* basic interval, the previous results can be extended to *some* integral equations with an *infinite* basic interval, for instance, to integral equations of the type

$$\phi(x) - \lambda \int_0^\infty K(x,y)\,\phi(y)\,dy = f(x), \tag{3}$$

or even of the type

$$\phi(x) - \lambda \int_{-\infty}^\infty K(x,y)\,\phi(y)\,dy = f(x), \tag{4}$$

or to the corresponding equations of the first kind.

To do this we consider any transformation of x and y which carries the basic interval into a *finite* one, for instance, in the case of equation (3), the transformation

$$x = \frac{\xi}{1-\xi}, \qquad y = \frac{\eta}{1-\eta}. \tag{5}$$

Of course, the kernel of the new integral equation may have some singularities, but if, in spite of this, it belongs to the class L_2, all goes through.

Furthermore, as in the case of the Volterra integral equation, we can always substitute for the Fredholm integral equation

$$\phi(x) - \lambda \int_0^1 K(x, y)\,\phi(y)\,dy = f(x) \tag{6}$$

a similar equation with the iterated kernel $K_n(x, y)$ as kernel. In fact, from (6) we can deduce (as in § 2·5) that

$$\int K(x, z)\,\phi(z)\,dz - \lambda \int K(x, z)\,dz \int K(z, y)\,\phi(y)\,dy = \int K(x, z)\,f(z)\,dz,$$

$$\frac{1}{\lambda}[\phi(x) - f(x)] - \lambda \int K_2(x, y)\,\phi(y)\,dy = \int K(x, y)\,f(y)\,dy,$$

that is,

$$\phi(x) - \lambda^2 \int K_2(x, y)\,\phi(y) = f(x) + \lambda \int K(x, y)\,f(y)\,dy \equiv f_2(x), \tag{7}$$

and successively we obtain

$$\left. \begin{aligned} \phi(x) - \lambda^3 \int K_3(x, y)\,\phi(y)\,dy = f_3(x) \equiv f_2(x) + \lambda \int K(x, y)\,f_2(y)\,dy, \\ \dots\dots\dots\dots\dots\dots\dots\dots\dots\dots\dots\dots\dots\dots\dots\dots\dots\dots\dots\dots\dots\dots\dots\dots \end{aligned} \right\} \tag{8}$$

This method can be used to eliminate some singularities of the kernel,† because iterated kernels are generally 'smoother' than the original kernel. For instance, if the given kernel is of the type

$$K(x, y) = \frac{H(x, y)}{|x - y|^\alpha} \quad (0 < \alpha < 1,\ H \text{ bounded}), \tag{9}$$

then it is easy to see that $K_n(x, y)$ belongs to the same type, but, since the number α is changed into the number $1 - n(1 - \alpha)$, which becomes *negative* if n is large enough, K_n becomes *bounded*.

In fact, if

$$K(x, y) = \frac{H(x, y)}{|x - y|^\alpha}, \quad K'(x, y) = \frac{H'(x, y)}{|x - y|^{\alpha'}},$$

† Even singularities which imply that $K(x, y)$ is not quadratically integrable!

and if $x < y$ (an unessential hypothesis), then

$$\int_0^1 K(x,z)\,K'(z,y)\,dz = \int_0^x \frac{H(x,z)\,H'(z,y)}{(x-z)^\alpha (y-z)^{\alpha'}}\,dz$$

$$+ \int_x^y \frac{H(x,z)\,H'(z,y)}{(z-x)^\alpha (y-z)^{\alpha'}}\,dz + \int_y^1 \frac{H(x,z)\,H'(z,y)}{(z-x)^\alpha (z-y)^{\alpha'}}\,dz,$$

and all three integrals on the right-hand side are functions of the type

$$\frac{H^*(x,y)}{|x-y|^{\alpha+\alpha'-1}} \quad (H^* \text{ bounded}),$$

as we can see by using the three substitutions

$$z = x - (y-x)\,t, \quad z = x + (y-x)\,t, \quad z = y + (y-x)\,t,$$

respectively.

Another important fact is that the fundamental properties of Fredholm equations are independent of the number of dimensions of the space (range or two-dimensional domain, etc.) of the variables x and y.

Hence the earlier theory can be easily extended to integral equations of the form

$$\phi(P) - \lambda \int_{E_n} K(P,Q)\,\phi(Q)\,dV_Q = f(P) \quad (P \in E_n), \tag{10}$$

where $P \equiv (x_1, x_2, \ldots, x_n)$ and $Q \equiv (y_1, y_2, \ldots, y_n)$ are two points of a fixed n-dimensional manifold E_n whose volume element (around Q) we designate by dV_Q. Moreover, if we write $\phi(x)$, $K(x,y)$, etc., instead of $\phi(P)$, $K(P,Q)$, etc., as some authors do, even the form of most of the previous results remains unchanged. However, we shall not use this device because the over-simplified notations $\phi(x)$, $K(x,y)$, etc., may be a source of misunderstanding.

In particular, the basic Fredholm theorem and the consequent alternative theorem of §§ 2·3–2·4 remain valid for equation (10), provided that the kernel $K(P,Q)$ belongs to the class L_2, i.e. provided both functions

$$\left[\int_{E_n} K^2(P,Q)\,dV_Q\right]^{\frac{1}{2}} = A(P), \quad \left[\int_{E_n} K^2(P,Q)\,dV_P\right]^{\frac{1}{2}} = B(Q) \tag{11}$$

exist almost everywhere in E_n and their squares are integrable, i.e.

$$\| K \|^2 = \int_{E_n} A^2(P)\, dV_P = \int_{E_n} B^2(Q)\, dV_Q \leqslant N^2, \qquad (12)$$

where N is a positive constant, etc.

If the kernel is symmetric, i.e. if $K(P,Q) = K(Q,P)$, the important upper bound (3·11-5) for the first eigenvalue λ_1 remains valid, i.e. we have

$$|\lambda_1| \leqslant \| \phi \|^2 \Big/ \int_{E_n} \int_{E_n} K(P,Q)\, \phi(P)\, \phi(Q)\, dV_P\, dV_Q, \qquad (13)$$

where ϕ is any L_2-function.

3·18. Vibrations of a Membrane

As an example of the previous remarks we shall now treat the important problem of the vibrations of a *membrane*, a problem which can be reduced to a two-dimensional Fredholm integral equation of the second kind.

A *membrane*, the two-dimensional counterpart of a flexible string, is a thin elastic sheet with negligible stiffness† whose equilibrium shape is a plane sheet (the domain D of the xy-plane, Fig. 10).

We shall suppose the membrane to be *fixed* at the boundary s of the domain D, and subject to a uniform tension τ. This means that if we cut the (undisturbed) membrane along any regular curve σ (σ may coincide partly or entirely with a part of the contour s), then the mechanical action of the part of the membrane on one side of σ upon that part on the other side can be replaced by a system of normal forces (contained in the xy-plane and directed toward the first-named side of σ) such that the resulting force acting on any element $d\sigma$ of σ has intensity $\tau\, d\sigma$, where τ is a *constant*.

Methods similar to those of §§ 1·1 and 1·9 (i.e. influence functions, etc.) can be used to determine the deformation of the membrane under the action of given loads, or, more exactly, to determine the normal displacement $z = z(x,y)$ of a point $P \equiv (x,y)$ of the membrane if the load $p(\xi, \eta)\, d\xi\, d\eta$ acts (parallel to Oz) on its two-dimensional

---

† If the stiffness is not negligible, the system is called a *plate* instead of a membrane.

element $d\xi\,d\eta$ at the point $Q = (\xi, \eta)$ of D. However, in this case it is better to start from the differential equation of the membrane†

$$\Delta z = \frac{\partial^2 z}{\partial x^2} + \frac{\partial^2 z}{\partial y^2} = -\frac{1}{\tau} p(x, y), \tag{1}$$

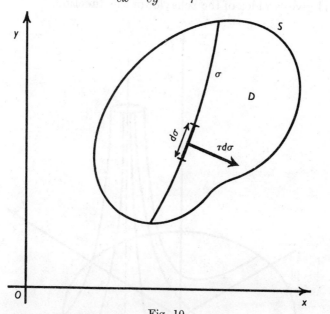

Fig. 10

because then we can utilize the results of § 2·7 on the Dirichlet problem. These results assure us that Green's function $G(x, y; \xi, \eta)$ of the domain D with respect to the operator Δ exists. It can be written

$$G(P, Q) \equiv G(x, y; \xi, \eta) = \log\frac{1}{r} - g(P, Q), \tag{2}$$

where

$$r = \sqrt{[(x - \xi)^2 + (y - \eta)^2]}$$

denotes the distance between the two points P and Q of D, and $g(P, Q)$ is the harmonic function inside D whose boundary values on s coincide with those of $\log(1/r)$ (so that $G(P, Q)$ vanishes on s). If D is a *circle* of radius a with its center at the origin O, then

$$G(P, Q) = \log\frac{1}{r} - \log\frac{1}{r_1}, \tag{3}$$

† See, for example, Frank-v. Mises [12], **2**, p. 340.

where r_1 is the distance of Q from the (exterior) point

$$P_1 \equiv \left(\frac{a^2 x}{x^2 + y^2}, \quad \frac{a^2 y}{x^2 + y^2} \right).$$

Fig. 11 gives an idea of the behavior of this function.

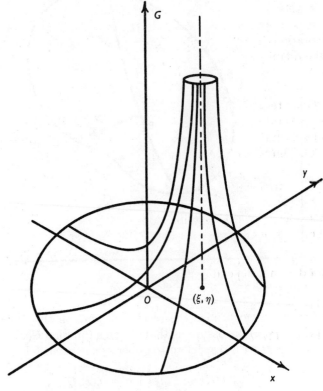

Fig. 11

With the help of Green's function, any twice-differentiable function $\phi(x, y)$ can be calculated, provided its boundary values on s are known as well as all the values of $\Delta\phi$ inside of D. We use the well-known *Green's theorem*† (or Green's third identity)

$$2\pi\phi(x, y) = \oint_s \frac{dG}{dn} \phi \, ds - \iint_D G(P, Q) \, \Delta\phi(Q) \, dS_Q. \tag{4}$$

† See, for example, Tricomi [45], p. 324.

Since the displacement $z(x, y)$ of the points of the membrane must vanish on the boundary s of D (where the membrane is fixed), from (4) and (1) we obtain (using obvious notation)

$$z(P) = \frac{1}{2\pi\tau} \iint_D G(P, Q)\, p(Q)\, dS_Q. \tag{5}$$

From this *static* equation we can obtain the corresponding *dynamic* equation governing the *free* vibrations of the membrane (i.e. vibrations subject to no exterior forces) by substituting for $p(Q)$ the quantity

$$-\mu(Q) \frac{\partial^2}{\partial t^2} z(Q),$$

where t denotes the time and $\mu(Q) \equiv \mu(\xi, \eta)$ is the *superficial density* of the (generally non-homogeneous) membrane, i.e. $\mu(\xi, \eta)\, d\xi\, d\eta$ coincides with the mass of the membrane element $d\xi\, d\eta$ at the point $Q = (\xi, \eta)$. Thus the dynamic equation is

$$z(P) = -\frac{1}{2\pi\tau} \iint_D G(P, Q)\, \mu(Q) \frac{\partial^2}{\partial t^2} z(Q)\, dS_Q. \tag{6}$$

In particular, for *harmonic* vibrations of angular frequency ω, i.e. for vibrations of the type

$$z = u(P)\, e^{\omega t i}, \tag{7}$$

where u does not depend on t, since

$$\frac{\partial^2 z}{\partial t^2} = -\omega^2 u(P)\, e^{\omega t i},$$

we obtain a *Fredholm homogeneous integral equation of the second kind*:

$$u(P) = \frac{\omega^2}{2\pi\tau} \iint_D G(P, Q)\, \mu(Q)\, u(Q)\, dS_Q. \tag{8}$$

This equation assumes the simpler form

$$u(P) - \lambda \iint_D G(P, Q)\, u(Q)\, dS_Q \quad \text{with} \quad \lambda = \frac{\omega^2 \mu}{2\pi\tau}, \tag{9}$$

if the membrane is *homogeneous*, i.e. if $\mu(Q) = \mu = \text{const.}$‡

---

‡ However, if $\mu(x, y)$ is not constant, we obtain an equation with the kernel

$$K(P, Q) = G(P, Q) \sqrt{[\mu(P)\mu(Q)]}$$

when we put $u(P) \sqrt{[\mu(P)]} = \phi(P)$ as in § 3·14.

The main property of the function $G(P,Q)$ as a kernel of an integral equation is its *symmetry*, i.e. $G(P,Q) = G(Q,P)$. This well-known property of Green's function can be easily deduced from the static equation (5), if use is made of *Betti's principle* (cf. § 1·1 and Tricomi [48]).

Another important property of the kernel $G(P,Q)$ is that it is *positive-definite* in the sense of § 3·12, i.e. if $\Phi(P)$ is any L_2-function (not vanishing almost everywhere), we have

$$J(\Phi, \Phi) = \iint_D \iint_D G(P,Q)\, \Phi(P)\, \Phi(Q)\, dS_P dS_Q > 0. \qquad (10)$$

This can be deduced from the static equation (5), which permits us to consider $J(\Phi, \Phi)$ as a *work* (cf. § 3·14), but it is more instructive to use the well-known identity

$$\iint_D \phi(P)\, \Delta\phi(P)\, dS_P + \iint_D \left[\left(\frac{\partial\phi}{\partial x}\right)^2 + \left(\frac{\partial\phi}{\partial y}\right)^2\right] dS_P + \oint_s \phi\, \frac{d\phi}{dn}\, ds = 0, \qquad (11)$$

which, for any twice-differentiable function $\phi(P)$ vanishing on the boundary s of D, gives us

$$\iint_D \phi(P)\, \Phi(P)\, dS_P = -\iint_D \left[\left(\frac{\partial\phi}{\partial x}\right)^2 + \left(\frac{\partial\phi}{\partial y}\right)^2\right] dS_P,$$

provided that $$\Phi \equiv \Delta\phi \qquad (12)$$

belongs to the class L_2. On the other hand, for such a function, it follows from (4) that

$$\phi(P) = -\frac{1}{2\pi} \iint_D G(P,Q)\, \Phi(Q)\, dS_Q; \qquad (13)$$

hence the previous identity can also be written

$$J(\Phi, \Phi) = 2\pi \iint_D \left[\left(\frac{\partial\phi}{\partial x}\right)^2 + \left(\frac{\partial\phi}{\partial y}\right)^2\right] dS_P > 0. \qquad (14)$$

In other words, we have identically

$$J(\Phi, \Phi) = 2\pi \iint |\operatorname{grad}\phi(P)|^2\, dS_P, \qquad (15)$$

provided that the function ϕ (vanishing on the boundary) is connected with Φ by means of the equivalent equations (12) or (13).

We can now conclude immediately that for the homogeneous (as well as for the non-homogeneous) membrane of any shape, there are an infinity of natural frequencies $\omega_1, \omega_2, \omega_3, \ldots$ which correspond to the infinite positive eigenvalues $\lambda_1, \lambda_2, \lambda_3, \ldots$ of the symmetric positive kernel $G(P, Q)$.

These eigenvalues can be evaluated approximately by using the methods of § 3·11, which can be immediately extended to the case of two (or more) dimensions. In particular, from (3·17-13) we obtain a useful upper bound for the smallest eigenvalue λ_1 of the kernel $G(P, Q)$. Namely,

$$\lambda_1 \leqslant \| \Delta\phi \|^2 \left[2\pi \iint_D |\operatorname{grad} \phi(P)|^2 \, dS_P \right]^{-1}, \tag{16}$$

where $\phi(P)$ is any twice-differentiable function which vanishes on the boundary s and is such that $\Delta\phi$ belongs to the class L_2.

Correspondingly, the first natural frequency ω_1 of a homogeneous membrane has the upper bound

$$\omega_1 \leqslant \sqrt{\left(\frac{\tau}{\mu}\right)} \| \Delta\phi \| \left[\iint |\operatorname{grad} \phi(P)|^2 \, dS_P \right]^{-\frac{1}{2}}. \tag{17}$$

Let us consider as an example the case of a *circular* homogeneous membrane of radius a. In this case the differential equation of the motion can be explicitly integrated by means of Bessel functions† and consequently the natural frequencies are given by the formula

$$\omega_{n, m} = \frac{1}{a} \sqrt{\left(\frac{\tau}{\mu}\right)} j_{n, m} \quad (m = 1, 2, 3, \ldots; n = 0, 1, 2, \ldots), \tag{18}$$

where $j_{n, m}$ denotes the mth positive zero (from the origin) of the Bessel function of the first kind $J_n(x)$. In particular, for the lowest frequency we have exactly

$$\omega_1 = \omega_{01} = \frac{1}{a} \sqrt{\left(\frac{\tau}{\mu}\right)} j_{0, 1} = \frac{1}{a} \sqrt{\left(\frac{\tau}{\mu}\right)} 2{\cdot}4048. \tag{19}$$

We can obtain an upper bound for this same frequency from (17) by putting

$$\phi(P) = a^2 - r^2, \tag{20}$$

where r denotes the distance of the point P from the center of the membrane. Since

$$\operatorname{grad}(r^2) = 2r \operatorname{grad} r, \quad |\operatorname{grad} \phi|^2 = 4r^2,$$

† See, for example, Frank-v. Mises [12], p. 348 et seq. or Morse [31], p. 150 et seq. and so on.

we obtain thus

$$\omega_1 \leqslant \sqrt{\left(\frac{\tau}{\mu}\right)} \left[16 \iint_D dS_P\right]^{\frac{1}{2}} \left[4 \iint_D \tau^2 dS_P\right]^{-\frac{1}{2}},$$

that is, $$\omega_1 \leqslant \frac{1}{a} \sqrt{\left(\frac{\tau}{\mu}\right)} 2\sqrt{2}, \qquad (21)$$

which is a good approximation to the exact value (19) since $2\sqrt{2} = 2 \cdot 8284$.

Furthermore, the method of integral equations can give us useful information, for example, about the variation of the natural frequencies when the shape of the membrane is modified slightly.†

---

† See Krall-Einaudi [28], **2**, pp. 265–7, who refers to some results of T. Boggio on the subject. See also Pólya-Szegö [38].

CHAPTER IV

Some Types of Singular or Non-Linear Integral Equations

4·1. Orientation and Examples

The material which will be considered in this last chapter is fragmentary. This is because of its very nature as well as the fact that we deal with theories still in the process of formation.† Two different aims of the research can already be distinguished:

(i) to determine whether or not a Fredholm theorem (and especially an alternative theorem) is valid for some classes of integral equations, even if the kernel does not belong to the class L_2;

(ii) to study typical classes of integral equations whose behavior deviates decidedly from that of the Fredholm equations.

We have given an idea of research of the first kind in §§ 1·12 and 3·17. We shall now consider some examples of research of the second kind.

In equations not governed by the Fredholm theorem three new phenomena often appear:

(1) the presence of finite accumulation points of the spectrum of eigenvalues, or even of a *continuous spectrum*, i.e. of eigenvalues filling whole segments of the λ-axis or even the entire λ-axis;

(2) the presence of eigenvalues of *infinite index*, i.e. of eigenvalues to which an infinite number of linearly independent eigenfunctions correspond;

(3) the presence of *bifurcation points* (in the real, non-linear case), i.e. of points of the λ-axis at each of which, as we pass from the left to the right, the number of solutions of the equation, although finite, changes.

---

† However, some parts are already quite well developed. See, for example, W. F. Trjitzinsky, 'General theory of singular integral equations with real kernels', *Trans. Amer. Math. Soc.* **46**, 1939, pp. 202–79 and N. I. Muskhelishvili [32].

These phenomena can already be observed in very simple integral equations for which the conditions of the previous chapters are not satisfied. As an example of an equation which has a continuous spectrum we shall consider the *Lalesco-Picard equation*

$$\phi(x) - \lambda \int_{-\infty}^{\infty} e^{-|x-y|} \phi(y)\, dy = f(x), \tag{1}$$

whose kernel has an infinite norm, since

$$\int_{-\infty}^{\infty} \int_{-\infty}^{\infty} e^{-2|x-y|}\, dx\, dy$$

$$= \frac{1}{2} \int_{-\infty}^{\infty} \{ e^{-2x} [e^{+2y}]_{-\infty}^{x} + e^{+2x} [e^{-2y}]_{x}^{\infty} \}\, dx = \int_{-\infty}^{\infty} dx.$$

If the functions $f(x)$ and $\phi(x)$ are both twice differentiable, our equation

$$\phi(x) - \lambda \left[e^{-x} \int_{-\infty}^{x} e^{y} \phi(y)\, dy + e^{x} \int_{x}^{\infty} e^{-y} \phi(y)\, dy \right] = f(x)$$

is substantially equivalent to the differential equation

$$\phi''(x) + 2\lambda \phi(x) - [\phi(x) - f(x)] = f''(x),$$

i.e. to the equation

$$\phi''(x) + (2\lambda - 1)\, \phi(x) = f''(x) - f(x).$$

Consequently for the case $f(x) \equiv 0$ we have

$$\phi(x) = A e^{\mu x} + B e^{-\mu x}, \tag{2}$$

where A and B are arbitrary constants and

$$\mu = \sqrt{(1 - 2\lambda)}, \tag{3}$$

provided the integral in equation (1) exists, i.e. provided $|\mathscr{R}\mu| < 1$, which, for λ real, implies $\lambda > 0$.

This shows that for the field of real numbers *the spectrum of equation* (1) *covers the infinite segment* $0 < \lambda < \infty$. Each point of this segment is an eigenvalue of index 2 of our equation. However, the corresponding eigenfunctions (2) do *not* belong to the class L_2 in $(-\infty, \infty)$.

More general equations of the form

$$\phi(x) - \lambda \int_{-\infty}^{\infty} k(x - y)\, \phi(y)\, dy = 0$$

(or similar ones with the integral ranging over $0, \infty$)

where $k(t)$ vanishes exponentially as $|t| \to \infty$, have been studied by Wiener and Hopf.† Recently S. Chandrasekhar has found important applications for these equations to the problem of radioactive equilibrium of a stellar atmosphere.‡

An important example of an integral equation with *eigenvalues of infinite index* is the equation with a '*Hankel kernel*' §

$$\phi(t) - \lambda \int_0^\infty J_\nu[2\sqrt{(tu)}]\,\phi(u)\,du = f(t), \qquad (4)$$

where $J_\nu(z)$ denotes the Bessel function of the first kind of order ν.

Many years ago I showed that¶ if $\nu > -1$, and if the Laplace transform

$$\psi(x) \equiv \mathscr{L}_x[t^{\frac{1}{2}\nu}\phi(t)] \equiv \int_0^\infty e^{-xt} t^{\frac{1}{2}\nu}\,\phi(t)\,dt \qquad (5)$$

exists, then, after multiplying both sides of (4) by $t^{\frac{1}{2}\nu}$, and applying the $\mathscr{L}$-transformation to both sides, the order of the infinite integrations in the double integral can be interchanged. Thus we obtain

$$\psi(x) - \lambda \int_0^\infty \phi(y)\,dy \int_0^\infty e^{-xt} t^{\frac{1}{2}\nu} J_\nu[2\sqrt{(ty)}]\,dt = g(x),$$

where
$$g(x) = \int_0^\infty e^{-xt} t^{\frac{1}{2}\nu} f(t)\,dt \equiv \mathscr{L}_x[t^{\frac{1}{2}\nu} f(t)]. \qquad (6)$$

In view of a well-known formula of the $\mathscr{L}$-transformation, we have

$$\int_0^\infty \phi(y)\,dy \int_0^\infty e^{-xt} t^{\frac{1}{2}\nu} J_\nu[2\sqrt{(ty)}]\,dt = x^{-(\nu+1)} \int_0^\infty e^{-y/x} y^{\frac{1}{2}\nu}\,\phi(y)\,dy$$

$$= x^{-(\nu+1)}\,\psi(1/x).$$

Hence the given equation assumes the 'algebraic' form

$$\psi(x) - \lambda x^{-(\nu+1)}\,\psi(1/x) = g(x). \qquad (7)$$

---

† N. Wiener-E. Hopf, 'Über eine Klasse singulärer Integralgleichungen', *Sitzungsber. Preuss. Akad. der Wiss.* 1931, p. 695. See also [34].

‡ Many papers in *Astrophys. J.* from **94**, 1941, on.

§ F. Tricomi, 'Autovalori e autofunzioni del nucleo di Hankel', *Atti Accad. Sci. Torino*, **71**, 1935–6, pp. 285–91.

¶ F. Tricomi, 'Sulla trasformazione e il teorema di reciprocità di Hankel', *Rend. Lincei* (6), **22**, 1935, pp. 564–71.

For the homogeneous case $f(x) \equiv g(x) \equiv 0$, we can set

$$x = e^{\xi}, \quad \psi(e^{\xi}) = \Psi(\xi) \tag{8}$$

and obtain the equation

$$\lambda \Psi(-\xi) = e^{(\nu+1)\xi} \Psi(\xi). \tag{9}$$

Applying this equation twice we obtain

$$\Psi(\xi) = \lambda^2 \Psi(\xi),$$

that is, $\lambda^2 = 1$. This shows that the only possible eigenvalues are $\lambda = \pm 1$.† Each of these eigenvalues is of *infinite index*, because for $\lambda = \pm 1$ equation (9) is satisfied by

$$\Psi(\xi) = \frac{E(\xi)}{1 + \lambda e^{(\nu+1)\xi}}, \tag{10}$$

where $E(x)$ denotes *any even* (analytic) function of x.

For example, if $\nu = 0$, $\lambda = 1$, $E(x) \equiv 1$, we obtain

$$\Psi(\xi) = \frac{1}{1 + e^{\xi}}, \quad \psi(x) = \frac{1}{1+x}, \quad \phi(x) = e^{-x},$$

and if $E(x) = 2\sqrt{\pi} \cosh(\tfrac{1}{2}x)$, we have

$$\Psi(\xi) = \sqrt{\pi}\, e^{-\frac{1}{2}\xi}, \quad \psi(x) = \sqrt{\frac{\pi}{x}}, \quad \phi(x) = \frac{1}{\sqrt{x}}.$$

Moreover, for $\nu = \pm \tfrac{1}{2}$ we can easily determine in this way all the *self-reciprocal functions* with respect to Fourier sine and cosine transformations.

Finally, in order to show an example of *bifurcation* points, we shall consider the very simple non-linear equation

$$\phi(x) - \lambda \int_0^1 \phi^2(y)\, dy = 1. \tag{11}$$

Set

$$\int_0^1 \phi^2(y)\, dy = \xi.$$

Then

$$\phi(x) = 1 + \lambda \xi, \tag{12}$$

and equation (11) becomes equivalent to the algebraic equation of the second degree
$$\xi = (1 + \lambda \xi)^2.$$

---

† This is, however, an immediate consequence of the *involutional* character of Hankel's transformation.

Hence we have $$\xi = \frac{1 - 2\lambda \pm \sqrt{(1 - 4\lambda)}}{2\lambda^2},$$

and $$\phi(x) = \frac{1 \pm \sqrt{(1 - 4\lambda)}}{2\lambda}. \tag{13}$$

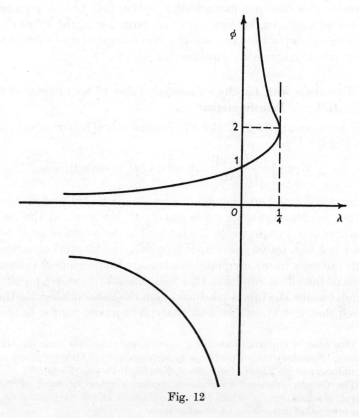

Fig. 12

Thus our equation has real solutions if and only if $\lambda \leqslant \frac{1}{4}$. It has exactly *two* solutions except for $\lambda = \frac{1}{4}$, when it has the (double) solution $\phi(x) \equiv 2$. For $\lambda = 0$ one solution is $\phi(x) \equiv 1$, while the second is *infinite*. This behavior is clearly shown by Fig. 12. The point $\lambda = \frac{1}{4}$ is a *bifurcation point*, while the point $\lambda = 0$ is a *singular point* of the equation.

The corresponding homogeneous equation

$$\phi(x) - \lambda \int_0^1 \phi^2(y)\, dy = 0 \tag{14}$$

always has the non-trivial solution

$$\phi(x) = \frac{1}{\lambda}. \tag{15}$$

However, this does not mean that equation (11) has an infinite number of solutions, because now the existence of the solutions (15) does not imply the existence of an infinite number of solutions of the 'non-homogeneous' equation (cf. also § 4·5).

4·2. Equations with Cauchy's Principal Value of an Integral and Hilbert's Transformation

We have seen in § 3·17 that a Fredholm integral equation with a kernel of the type

$$K(x, y) = \frac{H(x, y)}{|y - x|^\alpha} \quad (0 < \alpha < 1, \ H \text{ bounded})$$

can be transformed into a similar one with a bounded kernel. For this the hypothesis that $\alpha < 1$ is essential! However, in the important case $\alpha = 1$† (in which the integral of the equation must be considered as a *Cauchy principal integral*)‡ the integral equation differs radically from the equations considered in previous chapters.

One of the oldest results in this field consists of two reciprocity formulas which D. Hilbert § deduced from the Poisson integral. He showed that if $\Phi(z) = u + iv$ is an analytic function, regular in the

---

† This case is important in aerodynamics, etc. See, for example, E. Reissner, 'Boundary values problems in aerodynamics of lifting surfaces in non-uniform motion', *Bull. Amer. Math. Soc.* **55**, 1949, pp. 825–50.

‡ The *Cauchy principal integral* (or, Cauchy's *principal value* of the integral) of a function $f(x)$ which becomes infinite at an *interior* point $x = x_0$ of the interval of integration (a, b) is the limit

$$\lim_{\epsilon \to 0} \left(\int_a^{x_0 - \epsilon} + \int_{x_0 + \epsilon}^b \right) f(x)\, dx,$$

where $\qquad 0 < \epsilon \leqslant \min(x_0 - a, \ b - x_0)$.

We shall denote this integral by placing an *asterisk* over the usual integration sign.

If $f(x) = g(x)/(x - x_0)$, where $g(x)$ is any integrable function (in the sense of Lebesgue), then the above limit exists and is finite for almost every x_0 in (a, b); and if $g(x)$ belongs to the class L_p with $p > 1$, the principal integral also belongs to L_p. See, for example, Titchmarsh [43], pp. 144, 132.

§ *Göttinger Nachr.* 1904, pp. 213–59; Grundzüge usw. [18], p. 75.

circle $|z| < 1$, and $u(\theta)$ and $v(\theta)$ denote the real and imaginary parts of $\phi(z)$ on the boundary $|z| = 1$, then

$$
\begin{aligned}
u(\theta) &= \frac{1}{2\pi} \int_{-\pi}^{*\pi} \cot g \frac{\phi - \theta}{2} v(\phi)\, d\phi, \\
v(\theta) &= -\frac{1}{2\pi} \int_{-\pi}^{*\pi} \cot g \frac{\phi - \theta}{2} u(\phi)\, d\phi,
\end{aligned}
\tag{1}
$$

provided that
$$
\int_{-\pi}^{\pi} u(\theta)\, d\theta = \int_{-\pi}^{\pi} v(\theta)\, d\theta = 0.
\tag{2}
$$

For instance, if $\Phi(z) = z^n$ $(n = 1, 2, 3, \ldots)$ we have the formulae

$$
\begin{aligned}
\cos n\theta &= \frac{1}{2\pi} \int_{-\pi}^{*\pi} \cot g \frac{\phi - \theta}{2} \sin n\phi\, d\phi, \\
\sin n\theta &= -\frac{1}{2\pi} \int_{-\pi}^{*\pi} \cot g \frac{\phi - \theta}{2} \cos n\phi\, d\phi,
\end{aligned}
\tag{3}
$$

which are sometimes useful in changing a sine series into a cosine series and vice versa.

Instead of discussing the exact conditions for the validity of formulae (1), we shall study substantially equivalent questions concerning the values on the real axis of an analytic function $\Phi(z)$ regular inside the *half-plane* $\mathscr{I}z > 0$. In this case the Poisson integral for a harmonic function:

$$
f(x, y) = \frac{y}{\pi} \int_{-\infty}^{\infty} \frac{f(\xi, 0)}{(\xi - x)^2 + y^2}\, d\xi
$$

leads, under suitable conditions, to the simpler reciprocity formulae

$$
\begin{aligned}
u(x) &= \frac{1}{\pi} \int_{-\infty}^{*\infty} \frac{v(t)}{t - x}\, dt, \\
v(x) &= -\frac{1}{\pi} \int_{-\infty}^{*\infty} \frac{u(t)}{t - x}\, dt,
\end{aligned}
\tag{4}
$$

between the real part $u(x)$ and the imaginary part $v(x)$ of the boundary values $\Phi(x + i0)$ of the function $\Phi(z)$ on the real axis. In other words (v, u) and $(u, -v)$ are two pairs of 'conjugate functions' with respect to *Hilbert's transformation*

$$
f(x) = H_x[\phi(y)] \equiv \frac{1}{\pi} \int_{-\infty}^{*\infty} \frac{\phi(y)}{y - x}\, dy,
\tag{5}
$$

in the sense that the second is the transform of the first.

This transformation is discussed by Titchmarsh in his book [43] on Fourier integrals where the following three basic theorems are proved:†

THEOREM I. (Reciprocity theorem.) *If the function $\phi(x)$ belongs to the class L_p $(p > 1)$ in the basic interval $(-\infty, \infty)$, then formula (5) defines almost everywhere a function $f(x)$, which also belongs to L_p, whose Hilbert transform $H[f]$ coincides almost everywhere with $-\phi(x)$.*
That is, for any L_p-function

$$H\{H[\phi]\} = -\phi. \tag{6}$$

THEOREM II. (Generalized Parseval's formula.) *Let the functions $\phi_1(x)$ and $\phi_2(x)$ belong to the classes L_{p_1} and L_{p_2}, respectively. Then if*

$$\frac{1}{p_1} + \frac{1}{p_2} = 1, \tag{7}$$

we have $\displaystyle\int_{-\infty}^{\infty} \phi_1(x)\,\phi_2(x)\,dx = \int_{-\infty}^{\infty} H_x[\phi_1(y)]\,H_x[\phi_2(y)]\,dx.$ (8)

THEOREM III. *Let $\Phi(x+iy)$ be an analytic function, regular for $y > 0$, which, for all values of y, satisfies the condition*

$$\int_{-\infty}^{\infty} |\Phi(x+iy)|^p\,dx < K \quad (p > 1), \tag{9}$$

where K is a positive constant. Then as $y \to +0$, $\Phi(x+iy)$ converges for almost all x to a limit function:

$$\Phi(x+i0) = u(x) + iv(x)$$

whose real and imaginary parts $u(x)$ and $v(x)$ are two L_p-functions connected by the reciprocity formulae (4).

Hence, in particular, we have almost everywhere

$$\mathscr{R}\Phi(\xi+i0) = H_\xi[\mathscr{I}\Phi(\xi+i0)]. \tag{10}$$

Conversely, given any real function $v(x)$ of the class L_p, if we put

$$u(x) = H_x[v(y)], \tag{11}$$

† The theorems are stated here without proof. For the proofs see Titchmarsh [43], pp. 132, 138 and 139.

then the analytic function $\Phi(z)$ corresponding to the pair (u, v) can be calculated by means of the formula

$$\Phi(z) = \frac{1}{2\pi i} \int_{-\infty}^{\infty} \frac{u(t) + iv(t)}{t - z} dt \quad (\mathscr{I}z > 0), \tag{12}$$

and it satisfies condition (9).†

To these three theorems we add another, which plays a role similar to that of the *convolution theorem* in the theory of Laplace transformations.

THEOREM IV.‡ *Let the functions* $\phi_1(x)$ *and* $\phi_2(x)$ *belong to the classes* L_{p_1} *and* L_{p_2}, *respectively. Then if*

$$\frac{1}{p_1} + \frac{1}{p_2} < 1, \tag{13}$$

i.e. if $p_1 + p_2 < p_1 p_2$, *we have*

$$H\{\phi_1 H[\phi_2] + \phi_2 H[\phi_1]\} = H[\phi_1] H[\phi_2] - \phi_1 \phi_2 \tag{14}$$

almost everywhere.

To prove this, let

$$\phi_1 = v_1, \quad H[\phi_1] = u_1; \qquad \phi_2 = v_2, \quad H[\phi_2] = u_2.$$

Then the two analytic functions

$$\Phi_1(z) = \frac{1}{2\pi i} \int_{-1}^{1} \frac{u_1 + iv_1}{t - z} dt,$$

$$\Phi_2(z) = \frac{1}{2\pi i} \int_{-1}^{1} \frac{u_2 + iv_2}{t - z} dt$$

are regular for $\mathscr{I}z > 0$, and satisfy, respectively, the conditions

$$\int_{-\infty}^{\infty} |\Phi_1(x + iy)|^{p_1} dx < K_1,$$

$$\int_{-\infty}^{\infty} |\Phi_2(x + iy)|^{p_2} dx < K_2$$

---

† Hence we can be sure that an analytic function $\Phi(z)$, regular for $\mathscr{I}z > 0$, satisfies condition (9) whenever its limit values $\Phi(x + i0) = u(x) + iv(x)$ on the real axis exist almost everywhere and the functions $u(x)$ and $v(x)$ *both* belong to the class L_p with $p > 1$. In other words, under the previous hypothesis, it is not necessary *a priori* that u and v be connected by equations (4).

‡ F. G. Tricomi, 'On the finite Hilbert transformation', *Quart. J. Math.* (Oxford), (2) **2**, 1951, pp. 199–211.

for all values of y, where K_1 and K_2 are two suitable positive constants. Consequently, if we set $\Phi_1(z)\,\Phi_2(z) = \Phi(z)$, and further, in view of (13), set

$$\frac{1}{p_1} + \frac{1}{p_2} = \frac{1}{r} \quad (r > 1),$$

by virtue of Hölder's inequality,[†] we obtain

$$\int_{-\infty}^{\infty} |\,\Phi(x+iy)\,|^r\,dx < K_1^{r/p_1}\,K_2^{r/p_2}.$$

Hence the function $\Phi(z)$ satisfies condition (9) with $p = r > 1$. From (10) (i.e. from Theorem III), it follows that

$$u_1 u_2 - v_1 v_2 = \mathscr{R}\Phi(x+i0) = H[\mathscr{I}\,\Phi(x+i0)] = H[u_1 u_2 + v_1 v_2],$$

which is the *convolution theorem* (14) written in a different manner.

The 'skew-reciprocal' character (6) of Hilbert's transformation allows us to put Parseval's formula (8) into the form

$$\int_{-\infty}^{\infty} \phi_1(x)\,H_x[\phi_2(y)]\,dx = -\int_{-\infty}^{\infty} \phi_2(x)\,H_x[\phi_1(y)]\,dx. \tag{15}$$

After multiplying by π, this formula can be written

$$\int_{-\infty}^{\infty} \phi_1(x)\,dx \int_{-\infty}^{*\infty} \frac{\phi_2(y)}{y-x}\,dy = -\int_{-\infty}^{\infty} \phi_2(x)\,dx \int_{-\infty}^{*\infty} \frac{\phi_1(y)}{y-x}\,dy,$$

or, on interchanging x and y on the right-hand side,

$$\int_{-\infty}^{\infty} \phi_1(x)\,dx \int_{-\infty}^{*\infty} \frac{\phi_2(y)}{y-x}\,dy = \int_{-\infty}^{\infty} \phi_2(y)\,dy \int_{-\infty}^{*\infty} \frac{\phi_1(x)}{y-x}\,dx. \tag{16}$$

Thus we see that under suitable conditions an integration 'with star' can be interchanged with an ordinary integration. If the functions ϕ_1 and ϕ_2 vanish outside of a certain interval (a, b), then the previous formula becomes

$$\int_{a}^{b} \phi_1(x)\,dx \int_{a}^{*b} \frac{\phi_2(y)}{y-x}\,dy = \int_{a}^{b} \phi_2(y)\,dy \int_{a}^{*b} \frac{\phi_1(x)}{y-x}\,dx; \tag{17}$$

(17) *is valid provided* ϕ_1 *and* ϕ_2 *belong to the classes* L_{p_1} *and* L_{p_2} *respectively, with*

$$\frac{1}{p_1} + \frac{1}{p_2} = 1. \tag{18}$$

If the basic interval (a, b) *is finite,* (18) *can be replaced by the condition*

$$\frac{1}{p_1} + \frac{1}{p_2} \leqslant 1. \tag{18′}$$

† See Hardy, Littlewood and Pòlya [15], p. 140.

because in such an interval each L_p-function belongs also to the class $L_{p'}$ for $p' < p$; hence, to a pair p_1, p_2 satisfying (18), we can substitute a pair $p_1' \leqslant p_1$, $p_2' \leqslant p_2$ satisfying (18). More generally, by setting

$$\phi_1(x) \equiv 1, \quad \phi_2(x) = \Phi(x, x),$$

we can write

$$\int_a^b dx \int_a^{*b} \frac{\Phi(x, y)}{y - x} dy = \int_a^b dy \int_a^{*b} \frac{\Phi(x, y)}{y - x} dx, \tag{19}$$

provided the function $\Phi(x, x)$ belongs to the class L_p with $p > 1$ and the double integral

$$\int_a^b \int_a^b \frac{\Phi(x, y) - \Phi(x, x)}{x - y} dx \, dy \tag{20}$$

exists in the ordinary sense so that the classical Fubini theorem can be applied to it.†

In a similar manner, Theorem IV leads to the most useful particular case of the formula for the change of the order of two integrations *both 'with star'*.‡ For, since

$$\int_{-\infty}^{*\infty} \frac{\phi_1(x)}{x - x_0} dx \int_{-\infty}^{*\infty} \frac{\phi_2(y)}{y - x} dy - \int_{-\infty}^{*\infty} \phi_2(y) \, dy \int_{-\infty}^{*\infty} \frac{\phi_1(x)}{(x - x_0)(y - x)} dx$$

$$= \int_{-\infty}^{*\infty} \frac{\phi_1(x)}{x - x_0} dx \int_{-\infty}^{*\infty} \frac{\phi_2(y)}{y - x} dy$$

$$\qquad - \int_{-\infty}^{*\infty} \frac{\phi_2(y)}{y - x_0} dy \int_{-\infty}^{*\infty} \left(\frac{1}{x - x_0} - \frac{1}{x - y} \right) \phi_1(x) \, dx$$

$$= \pi^2 H_{x_0}\{\phi_1(x) \, H_x[\phi_2(y)]\} - \pi^2 H_{x_0}[\phi_2(y)] \, H_{x_0}[\phi_1(x)]$$

$$\qquad\qquad + \pi^2 H_{x_0}\{\phi_2(y) \, H_y[\phi_1(x)]\}$$

$$= \pi^2 H_{x_0}\{\phi_1(x) \, H_x[\phi_2(y)] + \phi_2(x) \, H_x[\phi_1(y)]\}$$

$$\qquad\qquad - \pi^2 H_{x_0}[\phi_1(y)] \, H_{x_0}[\phi_2(y)],$$

† For the details see the paper of the author quoted below.

‡ The general formula for changing the order of two Cauchy principal integrals can be written

$$\int_a^{*b} \frac{dz}{z - x} \int_a^{*b} \frac{F(x, y, z)}{y - z} dy = \int_a^{*b} dy \int_a^{*b} \frac{F(x, y, z)}{(z - x)(y - z)} dz - \pi^2 F(x, x, x).$$

Usually this formula is credited to H. Poincaré (1910, *Leçons de Méchanique Céleste*, 3, p. 254), but it can be found already in G. H. Hardy, 'The theory

from (14) it follows that almost everywhere

$$\int_{-\infty}^{*\infty} \frac{\phi_1(x)}{x-x_0}\, dx \int_{-\infty}^{*\infty} \frac{\phi_2(y)}{y-x}\, dy$$

$$= \int_{-\infty}^{*\infty} \phi_2(y)\, dy \int_{-\infty}^{*\infty} \frac{\phi_1(x)\, dx}{(x-x_0)\,(y-x)} - \pi^2 \phi_1(x_0)\, \phi_2(x_0), \quad (21)$$

provided that ϕ_1 and ϕ_2 belong to the classes L_{p_1} and L_{p_2} respectively, and that inequality (13) *is satisfied.*

In particular, if ϕ_1 and ϕ_2 vanish identically outside of (a, b), we have

$$\int_{a}^{*b} \frac{\phi_1(x)}{x-x_0}\, dx \int_{a}^{*b} \frac{\phi_2(y)}{y-x}\, dy$$

$$= \int_{a}^{*b} \phi_2(y)\, dy \int_{a}^{*b} \frac{\phi_1(x)\, dx}{(x-x_0)\,(y-x)} - \pi^2 \phi_1(x_0)\, \phi_2(x_0) \quad (a<x_0<b).$$

$$(22)$$

These considerations allow us to treat in L_p-space integral equations of the type indicated at the beginning of this section, provided the basic interval is the whole real axis $(-\infty, \infty)$.

In fact, consider first an equation of the *second kind*

$$\phi(x) - \lambda \int_{-\infty}^{*\infty} \frac{\phi(y)}{y-x}\, dy = f(x), \quad (23)$$

or, symbolically,

$$\phi(x) - \lambda \pi H_x[\phi(y)] = f(x).$$

Applying Hilbert's transformation to both sides of the equation and using (6) we have

$$H_x[\phi(y)] + \lambda \pi \phi(x) = H_x[f(y)].$$

Consequently $(1 + \lambda^2 \pi^2)\, \phi(x) = f(x) + \lambda \pi H_x[f(y)],$

of Cauchy's principal values' (fourth paper), *Proc. London Math. Soc.* (2), **7**, 1908, pp. 181–208. Hardy's proof is not very simple, and is similar to that which I later (without knowledge of his proof) gave in my paper on partial differential equations of *mixed type*; *Mem. Accad. Lincei Roma* (5), **14**, 1923, pp. 133–267—Russ. transl. Moscow, 1947, Engl. transl. 1948 (Brown University, U.S.A.).

For a simple proof of the previous formula under less restrictive hypotheses on the function F, see my recent paper: 'Sull'inversione dell'ordine di integrali "principali" nel senso di Cauchy', *Rend. Lincei* (8), **18**, 1955, p. 3–7.

or, written explicitly,

$$\phi(x) = \frac{1}{1 + \lambda^2 \pi^2} \left[f(x) + \lambda \int_{-\infty}^{*\infty} \frac{f(y)}{y - x} \, dy \right]. \tag{24}$$

Moreover, if the kernel is of the type

$$K(x, y) = \frac{1}{y - x} + K^*(x, y), \tag{25}$$

where K^* is bounded (or, at least, integrable), then by the previous treatment

$$(1 + \lambda^2 \pi^2) \, \phi(x) = f(x) + \lambda \int_{-\infty}^{*\infty} \frac{f(y)}{y - x} \, dy + \lambda \int_{-\infty}^{*\infty} K^*(x, y) \, \phi(y) \, dy$$

$$+ \lambda^2 \int_{-\infty}^{*\infty} \frac{dz}{z - x} \int_{-\infty}^{*\infty} K^*(x, y) \, \phi(y) \, dy.$$

If the order of the two successive integrations can be interchanged, this is an *ordinary*† Fredholm equation of the second kind with the kernel

$$K^*(x, y) + \lambda \int_{-\infty}^{*\infty} \frac{K^*(z, y)}{z - x} \, dz.$$

4·3. The Finite Hilbert Transformation and the Airfoil Equation

We shall now study equations with Cauchy's principal integrals over a *finite interval*. These equations have important applications, for example, in aerodynamics.

Of fundamental importance in this study is the *finite Hilbert transformation*

$$f(x) = \frac{1}{\pi} \int_{-1}^{*1} \frac{\phi(y)}{y - x} \, dy \equiv \mathscr{T}_x[\phi(y)], \tag{1}$$

where we assume the basic interval is $(-1, 1)$.

Until recently this transformation, in contrast to the infinite one, has received little attention.‡ A few of its properties can be deduced from the corresponding properties of the infinite transformation, by supposing that the function ϕ vanishes identically outside of the interval $(-1, 1)$.

---

† By *ordinary*, we mean that its integral is an ordinary one, not a Cauchy principal integral; however, the basic interval is still infinite.

‡ My recent paper 'On the finite Hilbert transformation', *Quart. J. Math.* (Oxford), (2), **2**, 1951, pp. 199–211, covers essentially the same material as this section.

For instance, from Parseval's formula (4·2-15),† we obtain

$$\int_{-1}^{1} \{\phi_1(x)\,\mathscr{T}_x[\phi_2(y)] + \phi_2(x)\,\mathscr{T}_x[\phi_1(y)]\}\,dx = 0, \tag{2}$$

provided that the functions ϕ_1 and ϕ_2 belong to the classes L_{p_1} and L_{p_2}, respectively (in the basic interval $(-1, 1)$), and that

$$\frac{1}{p_1} + \frac{1}{p_2} \leqslant 1. \tag{3}$$

Similarly, if the *equal* sign is excluded from this condition, we obtain from (4·2-14) the *convolution theorem*

$$\mathscr{T}\{\phi_1\mathscr{T}[\phi_2] + \phi_2\mathscr{T}[\phi_1]\} = \mathscr{T}[\phi_1]\,\mathscr{T}[\phi_2] - \phi_1\phi_2. \tag{4}$$

In other cases, however, the transformation $\mathscr{T}$ requires special treatment. For instance, the inversion formula (in the L_p-space)

$$\phi(x) = -\frac{1}{\pi}\int_{-\infty}^{*\infty} \frac{f(y)}{y-x}\,dy, \tag{5}$$

which can be immediately deduced from Theorem I of the previous section, is not satisfactory because its use requires knowledge of the function $f(x)$ outside of the basic interval $(-1, 1)$, where generally it does not vanish.

One difficulty in the study of the $\mathscr{T}$-transformation is that there exists no simple reciprocity theorem like Theorem I of the previous section; in fact, *the transformation has no unique inverse in the L_p-space $(p > 1)$.*

For instance, the $\mathscr{T}$-transform of the function

$$(1-x^2)^{-\frac{1}{2}}, \tag{6}$$

which belongs *to the class $2-0$,‡ vanishes identically* in the basic interval $(-1, 1)$; for, if we put $y = (1-t^2)/(1+t^2)$, we find

$$\mathscr{T}_x[(1-y^2)^{-\frac{1}{2}}] = 2\int_0^{*\infty} \frac{dt}{(1-x)-(1+x)\,t^2}$$
$$= (1-x^2)^{-\frac{1}{2}}\left[\log\left|\frac{\sqrt{(1-x)}+\sqrt{(1+x)}\,t}{\sqrt{(1-x)}-\sqrt{(1+x)}\,t}\right|\right]_0^{\infty} = 0. \tag{7}$$

† When applied to the finite transformation, *Parseval's formula* in its original form (4·2-8) is not particularly interesting.

‡ For the sake of brevity, we shall say, '*the function f belongs to the class p*', instead of 'f belongs to the class L_p' (in the basic interval considered); similarly, we may say '*the function f belongs to the class $r-0$*' instead of 'the function f belongs to the class L_p for all $1 < p < r$'. Similarly, we may say '*the function belongs to the class $r+0$*' if $f \in L_{r+\epsilon}$ for sufficiently small $\epsilon > 0$.

However, *outside of the interval* $(-1, 1)$, we have

$$\int_{-1}^{1} \frac{(1-y^2)^{-\frac{1}{2}}}{y-x} dy = -2(x^2-1)^{-\frac{1}{2}} \int_{0}^{\infty} \frac{d\tau}{1+\tau^2} = -\pi(x^2-1)^{-\frac{1}{2}}. \quad (8)$$

Note also that, as a consequence of (7), we obtain

$$\mathscr{T}_x[(1-y^2)^{\frac{1}{2}}] = \frac{1}{\pi}\int_{-1}^{*1} \frac{1-y^2}{\sqrt{(1-y^2)}} \frac{dy}{y-x} = -\frac{1}{\pi}\int_{-1}^{1} \frac{y+x}{\sqrt{(1-y^2)}} dy = -x. \quad (9)$$

The main problem for the transformation $\mathscr{T}$ is to find its inversion formula in the L_p-space $(p > 1)$, i.e. to solve the 'airfoil equation' (1)† by means of an L_p-function $(p > 1)$. It must, of course, be assumed that the given function $f(x)$ itself belongs to the class L_p.

To find an inversion formula we can use the convolution theorem (4), which can be applied to the function pair

$$\phi_1(x) \equiv \phi(x), \quad \phi_2(x) \equiv (1-x^2)^{\frac{1}{2}},$$

because the second function, being bounded, belongs to *any* class L_{p_2}, even for very large p_2. We thus obtain the equality

$$\mathscr{T}_x[-y\phi(y) + \sqrt{(1-y^2)}f(y)] = -xf(x) - \sqrt{(1-x^2)}\phi(x). \quad (10)$$

On the other hand, we have

$$\mathscr{T}_x[y\phi(y)] = \frac{1}{\pi}\int_{-1}^{*1} \frac{(y-x)+x}{y-x}\phi(y) dy = \frac{1}{\pi}\int_{-1}^{1} \phi(y) dy + xf(x).$$

Hence, from equation (10) it follows that *necessarily*

$$-\frac{1}{\pi}\int_{-1}^{1} \phi(y) dy + \mathscr{T}_x[\sqrt{(1-y^2)}f(y)] = -\sqrt{(1-x^2)}\phi(x),$$

that is, $$\sqrt{(1-x^2)}\phi(x) = -\mathscr{T}_x[\sqrt{(1-y^2)}f(y)] + C, \quad (11)$$

or, more explicitly,

$$\phi(x) = -\frac{1}{\pi}\int_{-1}^{*1} \sqrt{\left(\frac{1-y^2}{1-x^2}\right)} \frac{f(y)}{y-x} dy + \frac{C}{\sqrt{(1-x^2)}}. \quad (12)$$

---

† Equation (1) occurs in the theory of the motion of a lifting surface in an incompressible flow, and hence is often called the *airfoil equation*. It was studied among others by H. Söhngen (*Math. Zeitschr.* 45, 1939, pp. 245–64), who used an elegant method from potential theory (Poisson's integral), which, however, required strong restrictions on the given function. These restrictions were eliminated in a recent paper by the same author: 'Zur Theorie der endlichen Hilbert-Transformation', *Math. Zeitschr.* 60, 1954, pp. 31–51.

Here, in view of (7), the constant

$$C = \frac{1}{\pi} \int_{-1}^{1} \phi(y) \, dy \qquad (13)$$

has the character of an *arbitrary constant*.

The significance of the previous result is the following: *If the given equation* (1) *has any solution at all of the class* L_p ($p > 1$), *then this solution must have the form* (12).

Consequently, *the only non-trivial solutions of the class* L_p ($p > 1$) *of the homogeneous equation*

$$\frac{1}{\pi} \int_{-1}^{*1} \frac{\phi(y)}{y - x} \, dy = 0 \qquad (14)$$

are $C(1 - x^2)^{-\frac{1}{2}}$.

It remains to prove that the first term on the right-hand side of (12) satisfies equation (1). In order to do this, we show first that the term in question, i.e. *the function*

$$\phi_0(x) = \frac{1}{\pi} \int_{-1}^{1} \frac{\sqrt{(1-x^2)} - \sqrt{(1-y^2)}}{\sqrt{(1-x^2)}\,(y-x)} f(y) \, dy - \frac{1}{\pi} \int_{-1}^{*1} \frac{f(y)}{y - x} \, dy,$$

belongs to the class $\frac{4}{3} - 0$, *i.e. to any class* L_q *with* $1 < q < \frac{4}{3}$, *provided that the given function* $f(x)$ *belongs to the class* $\frac{4}{3} + 0$, *i.e. to a class* L_p *with* $p > \frac{4}{3}$.

Since

$$\phi_0(x) = \frac{1}{\pi} (1 - x^2)^{-\frac{1}{2}} \int_{-1}^{1} \frac{(x+y)f(y)}{\sqrt{(1-x^2)} + \sqrt{(1-y^2)}} \, dy - \mathscr{T}_x[f(y)],$$

the only difficulty is the determination of the class of the function

$$g(x) = \int_{-1}^{1} \frac{(x+y)f(y)}{\sqrt{(1-x^2)} + \sqrt{(1-y^2)}} \, dy,$$

because the factor $(1-x^2)^{-\frac{1}{2}}$ belongs to the class $2 - 0$, while $\mathscr{T}[f]$ (like $H[f]$) belongs to the same class as f.† In view of Hölder's inequality, we have

$$|g(x)|^{p'} \leqslant \int_{-1}^{1} \left[\frac{|x+y|}{\sqrt{(1-x^2)} + \sqrt{(1-y^2)}} \right]^{p'} dy \left[\int_{-1}^{1} |f(y)|^p \, dy \right]^{p'/p}$$

$$\left(p' = \frac{p}{p-1} \right),$$

† Titchmarsh [43], p. 132.

and consequently

$$\int_{-1}^{1} |g(x)|^{p'} dx$$
$$\leqslant \int_{-1}^{1} \int_{-1}^{1} \left[\frac{|x+y|}{\sqrt{(1-x^2)}+\sqrt{(1-y^2)}} \right]^{p'} dx\,dy \left[\int_{-1}^{1} |f(y)|^{p}\, dy \right]^{p'/p}.$$

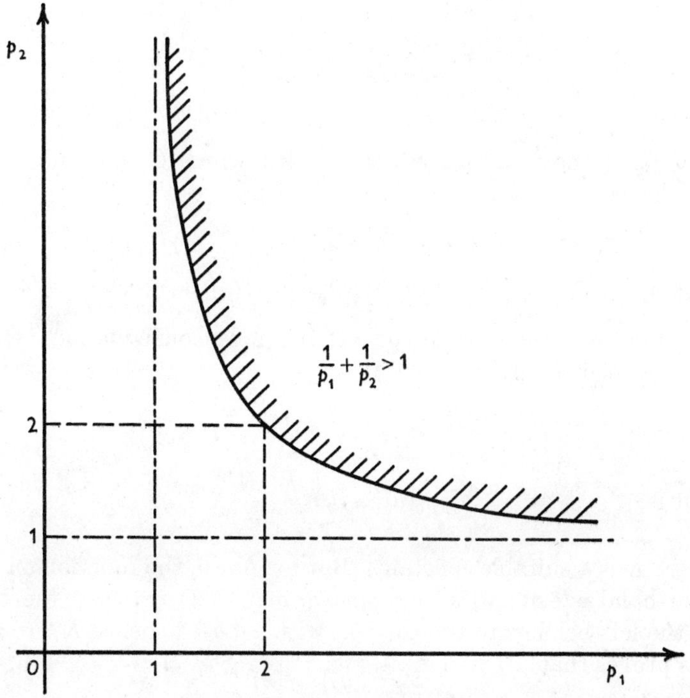

Fig. 13

Hence the function $g(x)$ belongs to the class $L_{p'}$ when the double integral on the right-hand side is finite, i.e. if $p' < 4$, because the integrand becomes infinite (of the order of $\frac{1}{2}p'$) only at the two boundary points $x = y = \pm 1$ of the square $(-1 \leqslant x \leqslant 1, -1 \leqslant y \leqslant 1)$. But $p' < 4$ if $p > \frac{4}{3}$; hence, remembering the rule for the class of the product of two functions,† we see that the product $(1-x^2)^{-\frac{1}{2}} g(x)$

† The product of two functions $f_1 \in L_{p_1}$ and $f_2 \in L_{p_2}$ is a function of class L_r with
$$\frac{1}{r} = \frac{1}{p_1} + \frac{1}{p_2}.$$

and thus the function $\phi_0(x)$ belong to the class L_q with $1 < q < \frac{4}{3}$ as long as $p > \frac{4}{3}$.

Now we apply the convolution theorem (4) to the pair

$$\phi_1(x) \equiv \sqrt{(1-x^2)}, \quad \phi_2(x) \equiv \phi_0(x),$$

and in a manner similar to that used to obtain (12), we get

$$\mathscr{T}_x\{\sqrt{(1-y^2)}\,\mathscr{T}_y[\phi_0(z)]\} = \frac{1}{\pi}\int_{-1}^{1}\phi_0(y)\,dy - \sqrt{(1-x^2)}\,\phi_0(x)$$

$$= C_0 - \sqrt{(1-x^2)}\,\phi_0(x).$$

But $\phi_0(x)$ may be expressed by (12) where C has the value C_0,† hence

$$\mathscr{T}_x\{\sqrt{(1-y^2)}\mathscr{T}_y[\phi_0(z)]\} = C_0 + \mathscr{T}_x[\sqrt{(1-y^2)}f(y)] - C_0,$$

that is, $$\mathscr{T}_x[\sqrt{(1-y^2)}\{\mathscr{T}_y[\phi_0(z)] - f(y)\}] = 0.$$

Using the previous result about the homogeneous equation (14), we see that necessarily

$$\sqrt{(1-x^2)}\,[\mathscr{T}_x[\phi_0(y)] - f(x)] = \frac{K}{\sqrt{(1-x^2)}},$$

that is, $$\mathscr{T}_x[\phi_0(y)] - f(x) = \frac{K}{1-x^2}, \qquad (15)$$

where K is a suitable constant. But for $K \neq 0$, the function on the right-hand side of (15) is *not summable* in $(-1, 1)$, while the function on the left belongs to the class L_q with $1 < q < \frac{4}{3}$; hence $K = 0$, and this proves that

$$\mathscr{T}_x[\phi_0(y)] \equiv f(x)$$

almost everywhere.

We have proved that:

If the given function $f(x)$ belongs to the class $\frac{4}{3} + 0$, then the airfoil equation (1) *has the solution*

$$\phi(x) = -\frac{1}{\pi}\int_{-1}^{*1}\sqrt{\left(\frac{1-y^2}{1-x^2}\right)}\frac{f(y)}{y-x}\,dy + \frac{C}{\sqrt{(1-x^2)}}, \qquad (16)$$

where C is an arbitrary constant, and the first term belongs at least to the class $\frac{4}{3} - 0$.

---

† This is permissible since $\mathscr{T}[(1-y^2)^{-\frac{1}{2}}] \equiv 0$; hence, the term with the constant C is without importance for the verification of formula (12).

Moreover, the second term, which belongs to the class $2 - 0$, *represents the most general solution of the corresponding homogeneous equation in the space* L_p $(p > 1)$.

If $f(x)$ belongs to a class L_p with $1 < p \leqslant \frac{4}{3}$, then we can no longer state that the function $\phi_0(x)$ certainly belongs to the class $1 + 0$; however, it remains true that

(i) if the given equation has any solution of the class $1 + 0$, then it must necessarily have the form (16),

(ii) if the function $\phi_0(x)$ belongs to the class $1 + 0$, then it necessarily satisfies the given equation.†

It is interesting to note further that, if the function $\phi(x)$ belongs to the class $2 + 0$, then its $\mathscr{T}$-transform $f(x)$ (which belongs also to the class $2 + 0$) necessarily satisfies the orthogonality condition

$$\int_{-1}^{1} (1 - x^2)^{-\frac{1}{2}} f(x) \, dx = 0. \tag{17}$$

In fact, under the previous hypothesis, we can apply the Parseval formula (2) to the pair of functions

$$\phi_1(x) \equiv (1 - x^2)^{-\frac{1}{2}}, \quad \phi_2(x) \equiv \phi(x),$$

and, in view of (7), obtain (17).

This allows us to put solution (16) into the alternate form

$$\phi(x) = -\frac{1}{\pi} \int_{-1}^{*1} \sqrt{\left(\frac{1 - x^2}{1 - y^2}\right)} \frac{f(y)}{y - x} \, dy + \frac{C'}{\sqrt{(1 - x^2)}}, \tag{18}$$

because, if (17) is valid, we also have

$$\int_{-1}^{*1} \left[\sqrt{\left(\frac{1 - x^2}{1 - y^2}\right)} - \sqrt{\left(\frac{1 - y^2}{1 - x^2}\right)}\right] \frac{f(y)}{y - x} \, dy$$

$$= \frac{1}{\sqrt{(1 - x^2)}} \int_{-1}^{*1} \frac{x + y}{\sqrt{(1 - y^2)}} f(y) \, dy = \frac{k}{\sqrt{(1 - x^2)}},$$

with

$$k = \int_{-1}^{1} \frac{y f(y)}{\sqrt{(1 - y^2)}} \, dy.$$

† The second paper of Söhngen, quoted at the beginning of this section, eliminates every doubt about the validity of (16), even if $f(x)$ belongs to a class L_p with $1 < p \leqslant \frac{4}{3}$. In this paper the author 'algebraized' the finite Hilbert transformation, with the help of another suitable functional transformation connected with the following formula (24).

Moreover, independent of condition (17), if at least one of the two functions

$$\frac{f(x)}{\sqrt{(1+x)}}, \quad \frac{f(x)}{\sqrt{(1-x)}}$$

is summable, then in view of the identities

$$\sqrt{\left(\frac{1-y^2}{1-x^2}\right)} = \sqrt{\left(\frac{1+x}{1-x}\right)} \sqrt{\left(\frac{1-y}{1+y}\right)} \left(1 + \frac{y-x}{1+x}\right)$$

$$= \sqrt{\left(\frac{1-x}{1+x}\right)} \sqrt{\left(\frac{1+y}{1-y}\right)} \left(1 - \frac{y-x}{1-x}\right)$$

solution (16) can be put into the two further alternate forms

$$\left.\begin{aligned}
\phi(x) &= -\frac{1}{\pi} \sqrt{\left(\frac{1+x}{1-x}\right)} \int_{-1}^{*1} \sqrt{\left(\frac{1-y}{1+y}\right)} \frac{f(y)}{y-x} dy + \frac{C''}{\sqrt{(1-x^2)}}, \\
\phi(x) &= -\frac{1}{\pi} \sqrt{\left(\frac{1-x}{1+x}\right)} \int_{-1}^{*1} \sqrt{\left(\frac{1+y}{1-y}\right)} \frac{f(y)}{y-x} dy + \frac{C'''}{\sqrt{(1-x^2)}}.
\end{aligned}\right\} \quad (19)$$

Some authors† use trigonometric series to solve the airfoil equation. This method is theoretically much less satisfying than the present one; however, it may be useful in practice. It is centered around the formula

$$\int_0^{*\pi} \frac{\cos(n\eta)\,d\eta}{\cos\eta - \cos\xi} = \pi \frac{\sin(n\xi)}{\sin\xi} \quad (n = 0, 1, 2, \ldots), \quad (20)$$

which, together with a similar one,

$$\int_0^{*\pi} \frac{\sin(n+1)\eta \sin\eta}{\cos\eta - \cos\xi}\,d\eta = -\pi \cos(n+1)\xi, \quad (21)$$

shows that the $\mathcal{T}$-transformation operates in a particularly simple manner on the Tchebichef polynomials‡

$$T_n(\cos\xi) = \cos(n\xi), \quad U_n(\cos\xi) = \frac{\sin(n+1)\xi}{\sin\xi}.$$

To be precise, we have

$$\mathcal{T}_x[(1-y^2)^{-\frac{1}{2}} T_n(y)] = U_{n-1}(x) \quad (n = 1, 2, 3, \ldots). \quad (22)$$

Hence, by expanding $f(x)$ in a series of polynomials $U_n(x)$, we can immediately deduce (at least formally) a corresponding expansion

---

† See, for example, G. Hamel [14] (1st ed.), p. 145.
‡ Szegö [42], p. 28 or Tricomi [49], p. 186.

of $\phi(y)$ in a series of polynomials $T_n(y)$, if we neglect the factor $(1-y^2)^{-\frac{1}{2}}$.

Formula (22), as well as the similar one

$$\mathscr{T}_x[(1-y^2)^{\frac{1}{2}} U_{n-1}(y)] = -T_n(x), \tag{23}$$

can be readily proved with the help of Theorem III of the previous section, by starting with the analytic functions

$$\Phi(z) = -(1-z^2)^{-\frac{1}{2}} [z-\surd(1-z^2)]^n \quad \text{and} \quad \Phi(z) = [z-\surd(1-z^2)\,i]^n,$$

respectively.†

Using the same method we can also prove the important formula

$$\mathscr{T}_x\left[\left(\frac{1-y}{1+y}\right)^\alpha\right] = \text{cotg}\,(\alpha\pi)\left(\frac{1-x}{1+x}\right)^\alpha - \frac{1}{\sin\,(\alpha\pi)} \quad (0<|\alpha|<1). \tag{24}$$

We start with the analytic function

$$\Phi(z) = \left(\frac{z-1}{z+1}\right)^\alpha - 1, \tag{25}$$

which satisfies condition (4·2-9), since for $|z+1|>2$ we have

$$\Phi(z) = -\alpha\frac{2}{z+1} + \binom{\alpha}{2}\left(\frac{2}{z+1}\right)^2 - \binom{\alpha}{3}\left(\frac{2}{z+1}\right)^3 + \dots.$$

Equation (24) is then an immediate consequence of (4·2-10), because on the real axis the function (25) reduces to a function $\Phi(x+i0)$ which is *real* outside of $(-1, 1)$, and for $-1<x<1$ we have

$$\Phi(x+i0) = \left(\frac{1-x}{1+x}\right)^\alpha e^{i\alpha\pi} - 1 = \left(\frac{1-x}{1+x}\right)^\alpha \cos\,(\alpha\pi) - 1 + i\left(\frac{1-x}{1+x}\right)^\alpha \sin\,(\alpha\pi). \tag{26}$$

Formula (24) is interesting because it shows that in some cases a function $\phi(y)$ which becomes *infinite* like $A(1-y)^{-\alpha}$ or $A(1+y)^{-\alpha}$ $(0<\alpha<1)$ as $y\to\pm1$ is carried by the $\mathscr{T}$-transformation into a

† These formulae are particular cases of more general formulae which give the $\mathscr{T}$-transform of a Jacobi polynomial $P_n^{(\alpha,\,\beta)}(y)$ multiplied by its weight-function $(1-y)^\alpha (1+y)^\beta$ as a linear combination of the original function itself and the hypergeometric function $F(-n-\alpha-\beta, n+1; 1-\alpha; \frac{1}{2}(1-x))$, which reduces also to a polynomial if $\alpha+\beta$ is an integer larger than $-n$. For details, see my paper mentioned at the beginning of this section.

function with similar behavior, if we neglect the fact that A is replaced by $\pm A \cot g(\alpha\pi)$.

That this behavior is common is shown by the following *asymptotic theorem*:

Let $\phi(x)$ be an L_p-function $(p>1)$ which in a small neighborhood $(-1, -1+\delta)$ $(\delta>0)$ of the point $x=-1$ can be written in the form

$$\phi(x)=A(1+x)^{-\alpha}+\psi(x) \quad (0\leqslant\alpha<1), \qquad (27)$$

where A is a constant, and $\psi(x)$ vanishes at $x=-1$ and satisfies (uniformly) a Lipschitz condition of positive order ϵ, i.e.

$$|\psi(x)-\psi(x_0)|<K|x-x_0|^\epsilon. \qquad (28)$$

Then the $\mathcal{T}$-transform $f(x)$ of $\phi(x)$ has the asymptotic representation

$$f(x)=A\cot g(\alpha\pi)(1+x)^{-\alpha}+O(1) \quad (x\to-1) \qquad (29)$$

if $0<\alpha<1$, and the asymptotic representation

$$f(x)=-\frac{A}{\pi}\log(1+x)+O(1) \quad (x\to-1) \qquad (30)$$

if $\alpha=0$.

If the point $x=-1$ is replaced by the point $x=+1$, then all remains the same except that $\cot g(\alpha\pi)$ is changed into $-\cot g(\alpha\pi)$ and $-\log(1+x)$ is changed into $+\log(1-x)$.

For the proof, we first observe that if the transform $f(x)$ becomes infinite at $x=-1$, then its asymptotic behavior for $x\to-1$ depends entirely on the values of $\phi(x)$ in an arbitrarily small interval $(-1, -1+\delta)$ $(\delta>0)$. In fact, if for x outside of $(-1+\delta, 1)$ we put

$$f(x)\equiv\mathcal{T}_x[\phi(y)]=\frac{1}{\pi}\int_{-1}^{*\delta-1}\frac{\phi(y)}{y-x}\,dy+\frac{1}{\pi}\int_{\delta-1}^{1}\frac{\phi(y)}{y-x}\,dy,$$

then the second term represents an analytic function regular in the whole z-plane cut along the segment $(-1+\delta, 1)$ of the real axis; hence it is regular even in the neighborhood of $x=-1$.

Consequently we can assume that (27) is valid in the whole interval $(-1, 1)$. Consider, in this interval, $\phi(x)$ written in the form

$$\phi(x)=\frac{A}{2^\alpha}\left(\frac{1-x}{1+x}\right)^\alpha+\psi^*(x),$$

where the function

$$\psi^*(x) = \psi(x) + \frac{A}{(1+x)^\alpha}\left[1 - \left(\frac{1-x}{2}\right)^\alpha\right]$$

satisfies the same sort of conditions as $\psi(x)$, in particular, the vanishing at $x = -1$. For $0 < \alpha < 1$, in view of (24), we have

$$\mathscr{T}_x\left[\frac{A}{2^\alpha}\left(\frac{1-y}{1+y}\right)^\alpha\right] = \frac{A}{2^\alpha}\cot g\,(\alpha\pi)\left(\frac{1-x}{1+x}\right)^\alpha - \frac{1}{\sin(\alpha\pi)}$$

$$= A\cot g\,(\alpha\pi)\,(1+x)^{-\alpha} + O\,(1) \quad (x \to -1);$$

and, for $\alpha = 0$, we have

$$\mathscr{T}_x[A] = A\log\left(\frac{1-x}{1+x}\right) = -A\log(1+x) + O(1).$$

Hence, to establish (29) and (30) we only have to show that

$$\mathscr{T}_x[\psi^*(y)] = O(1) \quad (x \to -1).$$

But this is a consequence of (28), since we have

$$\pi\mathscr{T}_x[\psi^*(y)] = \psi^*(x)\log\frac{1-x}{1+x} + \int_{-1}^1 \frac{\psi^*(y) - \psi^*(x)}{y-x}\,dy$$

and

$$\left|\int_{-1}^1 \frac{\psi^*(y) - \psi^*(x)}{y-x}\,dy\right| < K\int_{-1}^1 |y-x|^{\epsilon-1}\,dy < \frac{K}{\epsilon}2^{\epsilon+1}$$

while

$$\left|\psi^*(x)\log\left(\frac{1-x}{1+x}\right)\right| = |\psi^*(x) - \psi^*(-1)| \cdot \left|\log\left(\frac{1-x}{1+x}\right)\right|$$

$$< K(1+x)^\epsilon\,[\log(1-x) - \log(1+x)].$$

The passage from the case $x \to -1$ to the case $x \to 1$ offers no difficulties. It is only necessary to replace α in (24) by $-\alpha$.

In addition to the finite Hilbert transformation $\mathscr{T}$, it is often useful to consider the related transformation

$$F(x) = \frac{1}{\pi}\int_{-1}^1 \frac{\phi(t)}{t-z}\,dt, \tag{31}$$

where z is generally a complex number lying *outside* the segment $(-1, 1)$ on the real axis. This transformation changes a real function $\phi(t)$ of the class L_p $(p > 1)$ into a single-valued *analytic function* which is regular in the whole z-plane *cut along the segment* $(-1, 1)$ *of the real axis*, vanishes at infinity, and satisfies condition (4·2-9).

Transformation (31) is not essentially different from the well-known *Stieltjes transformation*.† To be precise, if we set

$$t = \frac{1-\tau}{1+\tau}, \quad z = \frac{1+s}{1-s}, \quad \frac{1}{1+\tau}\phi\left(\frac{1-\tau}{1+\tau}\right) = \psi(\tau), \quad -\frac{\pi}{1-s}F\left(\frac{1+s}{1-s}\right) = \Psi(s),$$

$$(32)$$

then equality (31) becomes

$$\Psi(s) = \int_0^\infty \frac{\psi(\tau)}{s+\tau}\, d\tau. \tag{33}$$

Thus transformation (31) is carried into that of Stieltjes and may therefore be inverted by the complex formula‡

$$\tfrac{1}{2}[\phi(x+0) + \phi(x-0)] = \frac{1}{2i}[F(x+i0) - F(x-i0)], \tag{34}$$

where, as usual,

$$\phi(x\pm 0) = \lim_{\epsilon\to+0}\phi(x\pm\epsilon), \quad F(x\pm i0) = \lim_{\epsilon\to+0}F(x\pm i\epsilon),$$

provided that both limits $\phi(x\pm 0)$ exist.

In other words, since $F(x+i0)$ and $F(x-i0)$ are conjugate complex numbers for ϕ real, we can state that

$$\mathscr{I}F(x+i0) = \begin{cases} \phi^*(x) & (-1 < x < 1), \\ 0 & (x < -1 \quad \text{or} \quad x > 1), \end{cases} \tag{35}$$

where

$$\phi^*(x) = \tfrac{1}{2}[\phi(x+0) + \phi(x-0)]. \tag{36}$$

But, by virtue of Theorem III of the previous section, we have §

$$\mathscr{R}F(x+i0) = H_x[\mathscr{I}F(y+i0)] = \mathscr{T}_x[\phi^*(y)] = \mathscr{T}_x[\phi(y)];$$

hence,

$$F(x+i0) = \mathscr{T}_x[\phi(y)] + i\phi^*(x) \quad (-1 < x < 1). \tag{37}$$

This formula, which for ϕ continuous can also be written as

$$F(x\pm i0) = \frac{1}{\pi}\int_{-1}^{*1}\frac{\phi(y)}{y-x}\,dy \pm i\phi(x), \tag{38}$$

is the key to the Carleman method of the next section.

An important consequence of the connection between (31) and the Stieltjes transformation is the possibility of utilizing well-

---

† See D. V. Widder [55], Chap. VIII.

‡ Widder [55], p. 340, Th. 76, where even the case $p = 1$ is admitted.

§ Remember that the set of points where $\phi^*(x) \neq \phi(x)$ is necessarily of measure *zero*. See, for example, Goffman (op. cit. in § 2·1), p. 195, Th. 6.

known properties of the Stieltjes transformation for the study of the function $F(x)$ on the real axis *outside of the interval* $(-1, 1)$.

For instance, *since the Stieltjes transformation* (33) *carries a function of the class* L_p $(p > 1)$ *into a similar one,*[†] *and since*

$$\int_0^\infty |\psi(\tau)|^p \, d\tau = 2^{1-p} \int_{-1}^1 |\phi(t)|^p (1+t)^{p-2} dt,$$

$$\int_0^\infty |\Psi(s)|^p \, ds = \pi^p 2^{1-p} \int_{-\infty}^{-1} + \int_1^\infty |F(z)|^p |1+z|^{p-2} dz,$$

we see that if the function $\phi(t)(1+t)^{1-2/p}$ *belongs to the class* L_p *in* $(-1, 1)$, *then the function* $F(x)(1+x)^{1-2/p}$ *belongs to the same class* L_p *in the double interval* $(-\infty, -1) + (1, \infty)$. In particular, when $\phi(t)$ belongs to the class L_2, so does $F(x)$.

4·4. Singular Equations of the Carleman Type[‡]

Given a singular integral equation of the second kind with the kernel

$$K(x, y) = \frac{H(x, y)}{y - x}, \tag{1}$$

we can often write

$$H(x, y) = H(x, x) + (y - x) H_y'(x, x) + \frac{1}{2!}(y - x) H_{yy}''(x, x) + \dots,$$

so that

$$K(x, y) = \frac{H(x, x)}{y - x} + K^*(x, y), \tag{2}$$

where $K^*(x, y)$ is bounded. Consequently the main problem in studying integral equations with kernels of type (2) is solving the *standard equation*

$$a(x)\phi(x) - \lambda \int_{-1}^{*1} \frac{\phi(y)}{y - x} dy = f(x), \tag{3}$$

where

$$a(x) = \frac{1}{H(x, x)}. \tag{4}$$

[†] This is an immediate consequence of the formula

$$\int_0^\infty |\Psi(s)|^p \, ds \leqslant \left[\frac{\pi}{\sin(\pi/p)}\right]^p \int_0^\infty |\Psi(\tau)|^p \, d\tau \quad (p > 1),$$

which can be deduced from the more general formula given by Widder [55], p. 370, by putting $K(t, u) = (t + u)^{-1}$.

[‡] My recent paper 'Sulle equazioni integrali del tipo di Carleman', *Annali di Matem.* (4) **39**, 1955, pp. 229–44, covers essentially the same material as this section.

We shall call (3) an *equation of the Carleman type* because it was discussed in 1922 in a paper by T. Carleman,† who gave an elegant, explicit solution of this equation in closed form.

Carleman uses two different methods based on the theory of analytic functions to deal with equation (3) (and with a related equation containing the difference $\phi(y) - \phi(x)$ under the integral sign). We shall consider only the second method, which consists essentially in the 'algebraization' of equation (3) by means of the earlier transformation (4·3-31).‡ This method is given here only heuristically. It would be difficult to make all the steps in the following reasoning rigorous, especially if unnecessary restrictions (such as the hypothesis that the unknown function be continuous) were dropped. The results, however, are valid.

Carleman§ observed that, if we put

$$F(z) = \frac{1}{\pi} \int_{-1}^{1} \frac{\phi(y)}{y - z} \, dy, \tag{5}$$

in view of (4·3-34), we have

$$\left. \begin{aligned} F(x + i0) - F(x - i0) &= 2i\phi(x), \\ F(x + i0) + F(x - i0) &= \frac{2}{\pi} \int_{-1}^{*1} \frac{\phi(y)}{y - x} \, dy, \end{aligned} \right\} \tag{6}$$

where we have assumed, for the sake of simplicity, that $\phi(x)$ is continuous in the open interval $(-1, 1)$. Consequently equation (3) assumes the 'algebraic' form

$$[a(x) - \lambda \pi i] \, F(x + i0) - [a(x) + \lambda \pi i] \, F(x - i0) = 2if(x). \tag{7}$$

---

† T. Carleman, 'Sur la résolution de certaines équations intégrales', *Arkiv for Mat., Astron. och Fysik*, **16**, 1922, 19 pp.

An equation of the Carleman type is thus an integral equation of the second kind with the singular kernel

$$\frac{1}{a(x)} \frac{1}{y - x}.$$

In the case of an equation of the *first* kind, it is not essentially different from the kernel $(y - x)^{-1}$ of the airfoil equation; however, in the case of an equation of the *second* kind, there are many differences.

‡ In a similar manner, equations of the Faltung type of § 1·8 were 'algebraized' by means of the Laplace transformation.

§ In order to keep the notation throughout this book consistent, we have sometimes modified Carleman's notation.

This equation can be simplified by setting

$$F(z) = e^{T(z)} U(z),\qquad(8)$$

provided that the function $T(z)$ satisfies the condition

$$[a(x) - \lambda\pi i] e^{T(x+i0)} = [a(x) + \lambda\pi i] e^{T(x-i0)}.\qquad(9)$$

We obtain

$$U(x+i0) - U(x-i0) = \frac{2if(x)}{a(x) - \lambda\pi i} e^{-T(x+i0)} = \frac{2if(x)}{a(x) + \lambda\pi i} e^{-T(x-i0)},$$

from which, if we consider the geometric mean of the two expressions for the difference on the left (which are equal), it follows that

$$U(x+i0) - U(x-i0)$$
$$= \frac{2if(x)}{\sqrt{[a^2(x) + \lambda^2\pi^2]}} \exp\{-\tfrac{1}{2}[T(x+i0) + T(x-i0)]\}.\qquad(10)$$

Now, in order to determine the function $T(z)$, we observe that from (9) we have

$$T(x+i0) - T(x-i0) = \log\frac{a(x) + \lambda\pi i}{a(x) - \lambda\pi i} = 2i\arctg\frac{\lambda\pi}{a(x)}.\qquad(11)$$

Consequently, although it is *not* necessary to do so, we *can* put

$$T(z) = \frac{1}{\pi}\int_{-1}^{1}\frac{\theta(t)}{t - z}\,dt,\qquad(12)$$

with

$$\theta(x) = \arctg_{(0,\,\pi)}\frac{\lambda\pi}{a(x)}.\qquad(13)$$

In fact, in view of (4·3-34), it follows from this that

$$\frac{1}{2i}[T(x+i0) - T(x-i0)] = \theta(x) = \arctg_{(0,\,\pi)}\frac{\lambda\pi}{a(x)},$$

in accordance with (11).

On the other hand, in view of (4·3-34), we have

$$\tfrac{1}{2}[T(x+i0) + T(x-i0)] = \frac{1}{\pi}\int_{-1}^{*1}\frac{\theta(t)}{t - x}\,dt;$$

hence, equation (10) now becomes

$$U(x+i0) - U(x-i0) = \frac{2i}{\sqrt{[a^2(x) + \lambda^2\pi^2]}} e^{-\tau(x)} f(x),$$

where
$$\tau(x) = \frac{1}{\pi} \int_{-1}^{*1} \frac{\theta(t)}{t-x} dt = \mathscr{T}_x[\theta(t)], \tag{14}$$

and *can* be satisfied by the function

$$U(z) = \frac{1}{\pi} \int_{-1}^{1} \frac{e^{-\tau(t)} f(t)}{\sqrt{[a^2(t) + \lambda^2 \pi^2]}} \frac{dt}{t-z}. \tag{15}$$

Finally, we determine ϕ by using the first equation (6) which, in view of (12) and (15), gives us

$$2i\phi(x) = e^{T(x+i0)} U(x+i0) - e^{T(x-i0)} U(x-i0)$$

$$= e^{\tau(x)+i\theta(x)} \left[\frac{1}{\pi} \int_{-1}^{*1} \frac{e^{-\tau(y)} f(y)}{\sqrt{[a^2(y) + \lambda^2 \pi^2]}} \frac{dy}{y-x} + i \frac{e^{-\tau(x)} f(x)}{\sqrt{[a^2(x) + \lambda^2 \pi^2]}} \right]$$

$$- e^{\tau(x)-i\theta(x)} \left[\frac{1}{\pi} \int_{-1}^{*1} \frac{e^{-\tau(y)} f(y)}{\sqrt{[a^2(y) + \lambda^2 \pi^2]}} \frac{dy}{y-x} - i \frac{e^{-\tau(x)} f(x)}{\sqrt{[a^2(x) + \lambda^2 \pi^2]}} \right].$$

After making some elementary transformations, we can write the solution in the form

$$\phi(x) = \frac{a(x) f(x)}{a^2(x) + \lambda^2 \pi^2} + \frac{\lambda e^{\tau(x)}}{\sqrt{[a^2(x) + \lambda^2 \pi^2]}} \int_{-1}^{*1} \frac{e^{-\tau(y)} f(y)}{\sqrt{[a^2(y) + \lambda^2 \pi^2]}} \frac{dy}{y-x}, \tag{16}$$

where, according to (14) and (13),

$$\tau(x) = \mathscr{T}_x[\theta(y)], \quad \theta(y) = \underset{(0,\,\pi)}{\operatorname{arctg}} \frac{\lambda \pi}{a(y)}.$$

We have presented Carleman's method only as a *heuristic* way of arriving at formula (16). We shall now obtain a similar formula (30′)† by using the methods of the previous section. We need first the $\mathscr{T}$-transform of the function

$$A(x) = \frac{e^{\tau(x)}}{\sqrt{[a^2(x) + \lambda^2 \pi^2]}} = \frac{\exp\left\{ \mathscr{T}_x\left[\underset{(0,\,\pi)}{\operatorname{arctg}} \left(\frac{\lambda \pi}{a(y)} \right) \right] \right\}}{\sqrt{[a^2(x) + \lambda^2 \pi^2]}}. \tag{17}$$

This can be deduced from Theorem III of § 4·2 and is itself of intrinsic interest.

To be precise, we shall show that

$$\lambda \pi \mathscr{T}_x[A(y)] = a(x) A(x) - \operatorname{sgn} \lambda \quad (\lambda \neq 0, \ -1 < x < 1), \tag{18}$$

supposing, for the sake of simplicity, that *the function $a(x)$ is continuous* in the basic interval $(-1, 1)$.

---

† Formula (16) lacks the last term of (30′), which contains the **arbitrary** constant C'.

For this we start with the analytic function

$$\Phi(z) = e^{T(z)} - 1,$$

where $T(z)$ is still given by (12). In view of (4·3-33) and the fact that $\theta(x)$ is continuous, even if $a(x)$ vanishes somewhere, $\Phi(z)$ reduces on the real axis to

$$\Phi(x+i0) = \begin{cases} \exp\left\{\frac{1}{\pi}\int_{-1}^{*1}\frac{\theta(t)}{t-x}dt + i\theta(x)\right\} - 1 & (-1 < x < 1), \\ \exp\left\{\frac{1}{\pi}\int_{-1}^{*1}\frac{\theta(t)}{t-x}dt\right\} - 1 & (x < -1,\ x > 1). \end{cases}$$

This shows that, for any positive p, both the real and imaginary parts of $\Phi(x+i0)$ are certainly L_p-functions in the interval $-1 < x < 1$. Outside of this interval, the imaginary part vanishes identically while the real part can be represented there by the series

$$\exp\left\{\frac{1}{\pi}\int_{-1}^{1}\frac{\theta(t)}{t-x}dt\right\} - 1 = \frac{M_0}{\pi x} + \frac{1}{\pi x^2}\left(M_1 + \frac{1}{2\pi}M_0^2\right) + \dots,$$

where $M_0, M_1, \dots$ are the successive *moments* of the function $\theta(x)$.†
Hence both the real and imaginary parts of $\Phi(x+i0)$ are of class L_p $(p > 1)$ on the whole real axis and consequently the function $\Phi(z)$ satisfies condition (4·2-9). Using again the symbol $\tau(x)$ for the function (14), from (4·2-10) it follows that

$$e^{\tau(x)}\cos\theta(x) - 1 = \mathscr{T}_x[e^{\tau(y)}\sin\theta(y)].$$

Now, remembering that $0 < \theta(x) < \pi$ or, even better, that

$$0 < \theta(x) < \tfrac{1}{2}\pi \quad (\lambda a(x) \geqslant 0), \qquad \tfrac{1}{2}\pi \leqslant \theta(x) < \pi \quad (\lambda a(x) \leqslant 0),$$

we notice that (if $\lambda \neq 0$) we also have

$$\sin\theta(x) = \frac{|\lambda|\pi}{\sqrt{[a^2(x) + \lambda^2\pi^2]}},$$

$$\cos\theta(x) = \pm\frac{a(x)}{\sqrt{[a^2(x) + \lambda^2\pi^2]}} = \frac{a(x)\operatorname{sgn}\lambda}{\sqrt{[a^2(x) + \lambda^2\pi^2]}},$$

† Namely, we put

$$M_n = \int_{-1}^{1}\theta(t)\,t^n\,dt \quad (n = 0, 1, 2, \dots).$$

Note that the constant term -1 must be present in order that the function $\Phi(x+i0)$ vanish at infinity.

where, as above, the radical is always considered positive. Hence, after multiplying by sgn λ, the previous equality can be written in the form

$$\frac{a(x)\,e^{\tau(x)}}{\sqrt{[a^2(x)+\lambda^2\pi^2]}} - \operatorname{sgn}\lambda = \lambda\pi\mathscr{T}_x\left[\frac{e^{\tau(y)}}{\sqrt{[a^2(y)+\lambda^2\pi^2]}}\right],$$

and this is exactly formula (18) which was to be proved.

If in the previous reasoning the function $\theta(x)$ is replaced by $\theta(x) - \pi$, then the function $\tau(x)$ is carried into

$$\tau(x) - \log\left(\frac{1-x}{1+x}\right),$$

and consequently $A(x)$ is carried into

$$\frac{1+x}{1-x}A(x),$$

while the signs of $\cos\theta(x)$ and $\sin\theta(x)$ are both changed. Hence, in addition to (18), we can write the further equality

$$\lambda\pi\mathscr{T}_x\left[\frac{1+y}{1-y}A(y)\right] = a(x)\frac{1+x}{1-x}A(x) + \operatorname{sgn}\lambda, \qquad (19)$$

and, by adding it to (18) and dividing by 2, we have

$$a(x)\frac{A(x)}{1-x} - \lambda\pi\mathscr{T}_x\left[\frac{A(y)}{1-y}\right] = 0. \qquad (20)$$

This last formula already yields an interesting result about the main problem of this section: *The homogeneous equation*

$$a(x)\,\phi(x) - \lambda\int_{-1}^{*1}\frac{\phi(y)}{y-x}\,dy = 0 \qquad (21)$$

always has the non-trivial solution

$$\phi_\lambda(x) = \frac{A(x)}{1-x}. \qquad (22)$$

Moreover, we can prove that: *If the function $a(x)$ is not only continuous (as already assumed), but also satisfies a Lipschitz condition of positive order ϵ in both the small intervals $(-1, -1+\delta)$ and $(1-\delta, 1)$ $(\delta > 0)$, then the function $\phi_\lambda(x)$, in spite of the divisor $1-x$, belongs to the class L_p $(p > 1)$.*

In fact, considering that for any pair x, x_0 of points in the interval $(-1, 1)$ we have

$$\theta(x) - \theta(x_0) = \operatorname*{arctg}_{(0,\,\pi)} \frac{\lambda\pi}{a(x)} - \operatorname*{arctg}_{(0,\,\pi)} \frac{\lambda\pi}{a(x_0)} = \operatorname*{arctg}_{(-\frac{1}{2}\pi,\,\frac{1}{2}\pi)} \frac{a(x_0)}{\lambda\pi} - \operatorname*{arctg}_{(-\frac{1}{2}\pi,\,\frac{1}{2}\pi)} \frac{a(x)}{\lambda\pi},$$

and consequently†

$$|\theta(x) - \theta(x_0)| < \left| \frac{a(x) - a(x_0)}{\lambda\pi} \right|,$$

our hypothesis about the function $a(x)$ allows us to apply the asymptotic theorem of the previous section to the function $\tau(x) = \mathscr{T}_x[\theta(y)]$. If we set

$$\alpha = \frac{1}{\pi}\theta(-1) = \frac{1}{\pi}\operatorname*{arctg}_{(0,\,\pi)} \frac{\lambda\pi}{a(-1)}, \quad \beta = \frac{1}{\pi}\theta(1) = \frac{1}{\pi}\operatorname*{arctg}_{(0,\,\pi)} \frac{\lambda\pi}{a(1)}, \quad (23)$$

we can write

$$\begin{aligned} \tau(x) &= -\alpha\log(1+x) + O(1) & (x \to -1), \\ \tau(x) &= \beta\log(1-x) + O(1) & (x \to 1). \end{aligned} \quad (24)$$

Hence we have
$$\begin{aligned} A(x) &= O[(1+x)^{-\alpha}] & (x \to -1), \\ A(x) &= O[(1-x)^{\beta}] & (x \to 1), \end{aligned}$$

and consequently

$$\begin{aligned} \phi_\lambda(x) &= O[(1+x)^{-\alpha}] & (x \to -1), \\ \phi_\lambda(x) &= O[(1-x)^{-(1-\beta)}] & (x \to 1). \end{aligned} \quad (25)$$

This shows that the function $\phi_\lambda(x)$ belongs to the class L_p for $p < 1/\gamma$, where
$$\gamma = \max(\alpha, 1-\beta). \quad (26)$$

It is important to notice further that the change of λ into $-\lambda$ carries $\theta(x)$ into $\pi - \theta(x)$ and $A(x)$ into

$$\frac{\exp\left\{\log\left(\dfrac{1-x}{1+x}\right) - \tau(x)\right\}}{\sqrt{[a^2(x) + \lambda^2\pi^2]}} = \frac{1-x}{1+x} A^*(x),$$

where
$$\frac{e^{-\tau(x)}}{\sqrt{[a^2(x) + \lambda^2\pi^2]}} = A^*(x). \quad (27)$$

Consequently we can say that the change of λ into $-\lambda$ carries the function
$$\frac{1+x}{1-x} A(x)$$

† Remember that $\dfrac{d}{dx}\operatorname{arctg} x = \dfrac{1}{1+x^2} \leqslant 1$.

into $A^*(x)$. It follows that

$$a(x)\,A^*(x) + \lambda\pi\mathscr{T}_x[A^*(y)] = \operatorname{sgn}\lambda. \tag{28}$$

Now, in order to verify rigorously Carleman's solution (16) and to show that functions of the type $C\phi_\lambda(x)$ are the only non-trivial solutions of the homogeneous equation (21), we start from the given equation (3), which 'composed' with the singular kernel

$$\frac{1-x}{\pi}\frac{A^*(x)}{y-x}$$

becomes

$$\mathscr{T}_x\{(1-y)\,A^*(y)\,[a(y)\,\phi(y) - f(y)]\}$$
$$= \frac{\lambda}{\pi}\int_{-1}^{*1}\frac{(1-z)\,A^*(z)}{z-x}\,dz\int_{-1}^{*1}\frac{\phi(y)}{y-z}\,dy.$$

The function $(1-x)\,A^*(x)$, *being bounded*,† belongs to the class $L_{p'}$, for any large p'. Hence, *if the unknown function ϕ belongs to any class L_p with $p > 1$*, formula (4·2-22) can be used to interchange the order of the two successive integrations 'with star' on the right-hand side. Hence we can also write

$$\mathscr{T}_x\{(1-y)\,A^*(y)\,[a(y)\,\phi(y) - f(y)]\}$$
$$= \frac{\lambda}{\pi}\int_{-1}^{*1}\frac{\phi(y)}{y-x}\,dy\int_{-1}^{*1}\left(\frac{1}{z-x} - \frac{1}{z-y}\right)(1-z)\,A^*(z)\,dz$$
$$\qquad\qquad\qquad\qquad - \lambda\pi(1-x)\,A^*(x)\,\phi(x)$$
$$= \lambda\int_{-1}^{*1}\{\mathscr{T}_x[(1-z)\,A^*(z)] - \mathscr{T}_y[(1-z)\,A^*(z)]\}\frac{\phi(y)}{y-x}\,dy$$
$$\qquad\qquad\qquad\qquad - \lambda\pi(1-x)\,A^*(x)\,\phi(x).$$

Now the $\mathscr{T}$-transform of the function $(1-x)\,A^*(x)$ can be easily evaluated with the help of formula (28). To be precise, we find

$$\mathscr{T}_x[(1-y)\,A^*(y)] = (1-x)\,\mathscr{T}_x[A^*(y)] - \frac{1}{\pi}\int_{-1}^{1}A^*(y)\,dy$$
$$= \frac{1-x}{\lambda\pi}[\operatorname{sgn}\lambda - a(x)\,A^*(x)] - \frac{1}{\pi}\int_{-1}^{1}A^*(y)\,dy$$

---

† In fact, the function is continuous in the open interval $(-1, 1)$ and, in view of (24), in the neighborhood of the end-points, we have

$$(1-x)\,A^*(x) = O[(1+x)^\alpha] \quad (x \to -1),$$
$$(1-x)\,A^*(x) = O[(1-x)^{1-\beta}] \quad (x \to 1).$$

and

$$\mathcal{T}_x[(1-z)\,A^*(z)] - \mathcal{T}_y[(1-z)\,A^*(z)]$$
$$= \frac{y-x}{\lambda\pi}\,\mathrm{sgn}\,\lambda + \frac{1}{\lambda\pi}\,[(1-y)\,a(y)\,A^*(y) - (1-x)\,a(x)\,A^*(x)].$$

Using this result, the previous equality becomes

$$\mathcal{T}_x\{(1-y)\,A^*(y)\,[a(y)\,\phi(y) - f(y)]\}$$
$$= \frac{\mathrm{sgn}\,\lambda}{\pi}\int_{-1}^{1}\phi(y)\,dy + \mathcal{T}_x[(1-y)\,a(y)\,A^*(y)\,\phi(y)]$$
$$- (1-x)\,a(x)\,A^*(x)\,\mathcal{T}_x[\phi(y)] - \lambda\pi(1-x)\,A^*(x)\,\phi(x).$$

Since the terms in the $\mathcal{T}$-transform of $(1-y)\,a(y)\,A^*(y)\,\phi(y)$ are *the same* on the left and on the right and the $\mathcal{T}$-transform of $\phi(y)$ can be eliminated by using equation (3), we have

$$-\mathcal{T}_x[(1-y)\,A^*(y)f(y)] = -\frac{1-x}{\lambda\pi}\,a(x)\,A^*(x)\,[a(x)\,\phi(x) - f(x)]$$
$$- \lambda\pi(1-x)\,A^*(x)\,\phi(x) + \frac{C}{\lambda\pi},$$

where C is an (arbitrary) constant. But we also have

$$-\frac{1-x}{\lambda\pi}\,a^2(x)\,A^*(x)\,\phi(x) - \lambda\pi(1-x)\,A^*(x)\,\phi(x)$$
$$= -\frac{1-x}{\lambda\pi}\,[a^2(x) + \lambda^2\pi^2]\,A^*(x)\,\phi(x) = -\frac{1-x}{\lambda\pi}\,\frac{\phi(x)}{A(x)};$$

hence we can conclude that from the given equation it follows necessarily that

$$\phi(x) = \frac{a(x)f(x)}{a^2(x) + \lambda^2\pi^2} + \lambda\pi\,\frac{A(x)}{1-x}\,\mathcal{T}_x[(1-y)\,A^*(y)f(y)] + C\,\frac{A(x)}{1-x}. \quad (29)$$

Even better, if the $\mathcal{T}$-transform of $A^*(y)f(y)$ exists,† we can write

$$\mathcal{T}_x[(1-y)\,A^*(y)f(y)] = (1-x)\,\mathcal{T}_x[A^*(y)f(y)] - \int_{-1}^{1}A^*(y)f(y)\,dy,$$

and the previous solution assumes Carleman's form

$$\phi(x) = \frac{a(x)f(x)}{a^2(x) + \lambda^2\pi^2} + \lambda\pi A(x)\,\mathcal{T}_x[A^*(y)f(y)] + C'\,\frac{A(x)}{1-x}. \quad (30)$$

† In view of the behavior of $A^*(y)$ at $y = \pm 1$, a sufficient condition for this existence is that $f(y)$ belong to a class L_q with $q > 1/(1-\beta)$.

or, more explicitly,

$$\phi(x) = \frac{a(x)f(x)}{a^2(x) + \lambda^2\pi^2} + \lambda\frac{e^{\tau(x)}}{\sqrt{[a^2(x) + \lambda^2\pi^2]}}\int_{-1}^{*1}\frac{e^{-\tau(y)}f(y)}{\sqrt{[a^2(y) + \lambda^2\pi^2]}}\frac{dy}{y - x}$$
$$+ C'\frac{e^{\tau(x)}}{(1 - x)\sqrt{[a^2(x) + \lambda^2\pi^2]}}, \qquad (30')$$

where C' denotes an arbitrary constant.

The significance of the previous result is the following: *If the given equation* (3) *has a solution* $\phi(x)$ *of the class* L_p $(p > 1)$, *it must necessarily have the form* (29)–(30). Consequently, we can already state that *the only non-trivial solutions of the class* L_p $(p > 1)$ *of the homogeneous equation* (21) *are of the form* $C\phi_\lambda(x)$, but it is yet far from certain that the function

$$F(x) \equiv \frac{a(x)f(x)}{a^2(x) + \lambda^2\pi^2} + \lambda\pi\frac{A(x)}{1 - x}\mathscr{T}_x[(1 - y)A^*(y)f(y)],$$

which can also be written as

$$F(x) \equiv \frac{a(x)f(x)}{a^2(x) + \lambda^2\pi^2} + \lambda\pi\phi_\lambda(x)\mathscr{T}_x[(1 - y)A^*(y)f(y)], \qquad (31)$$

actually satisfies the given equation (3).

To remove this doubt, we first observe that if the given function $f(x)$ belongs to the class L_q $(q > 1)$, then, since $(1 - x)A^*(x)$ *is bounded*, the functions

$$(1 - y)A^*(y)f(y) \quad\text{and}\quad \mathscr{T}_x[(1 - y)A^*(y)f(y)],$$

as well as the first term of $F(x)$, belong to the class L_q. Consequently, as already observed, since the function $\phi_\lambda(x)$ belongs to the class $(1/\gamma) - 0$ with the value (26) of γ, we can be certain that the function $F(x)$ belongs to the class L_{q^*} $(q^* > 1)$ provided that†

$$q^* < \frac{q}{1 + \gamma q}, \quad q > \frac{1}{1 - \gamma}. \qquad (32)$$

Secondly, we observe that if with a suitable value of the constant C, we put

$$\phi_0(x) = F(x) + C\phi_\lambda(x),$$

† For the sake of brevity, we shall not show that a better result could be obtained with a method similar to the one of the previous section.

and reverse the order of the steps which gave us formula (29), then we obtain the equality

$$\mathscr{T}_x\{(1-y)\,A^*(y)\,[a(y)\,\phi_0(y)-f(y)]\}$$
$$=\frac{\lambda}{\pi}\int_{-1}^{*1}\phi_0(y)\,dy\int_{-1}^{*1}\frac{(1-z)\,A^*(z)}{(z-x)(y-z)}\,dz-\lambda\pi(1-x)\,A^*(x)\,\phi_0(x).$$

But, since the function $(1-x)\,A^*(x)$ is bounded and the function $\phi_0(x)$ belongs† to a class L_{q^*} with $q^*>1$, formula (4·2-22) can be applied and the previous equality becomes

$$\mathscr{T}_x\{(1-y)\,A^*(y)\,[a(y)\,\phi_0(y)-f(y)]\}=\lambda\mathscr{T}_x\left[(1-z)\,A^*(z)\int_{-1}^{*1}\frac{\phi_0(y)}{y-z}\,dy\right],$$

that is,

$$\mathscr{T}_x\left\{(1-z)\,A^*(z)\left[a(z)\,\phi_0(z)-\lambda\int_{-1}^{*1}\frac{\phi_0(y)}{y-z}\,dy-f(z)\right]\right\}=0.$$

Using the results of the previous section, it follows that

$$a(x)\,\phi_0(x)-\lambda\int_{-1}^{*1}\frac{\phi_0(y)}{y-x}\,dy-f(x)=\frac{K}{(1-x)^{\frac{3}{2}}\,(1+x)^{\frac{1}{2}}\,A^*(x)},\qquad(33)$$

where K is a constant. For $K\neq0$, we have

$$\frac{K}{(1-x)^{\frac{3}{2}}\,(1+x)^{\frac{1}{2}}\,A^*(x)}=\begin{cases}O[(1+x)^{-(\alpha+\frac{1}{2})}] & (x\to-1),\\ O[(1-x)^{-(\frac{3}{2}-\beta)}] & (x\to1).\end{cases}$$

Hence for $\alpha\geqslant\frac{1}{2}$ or $\beta\leqslant\frac{1}{2}$ or both, K must necessarily vanish; otherwise, the function on the right-hand side of (33) belongs to no L_p class with $p\geqslant1$,‡ while the function on the left belongs to L_{q^*} with $q^*>1$. There remains the case $0<\alpha<\frac{1}{2}$, $\frac{1}{2}<\beta<1$ in which $\gamma>\frac{1}{2}$ and consequently the right-hand side of (33) is a function of class p with

$$p=\min\left(\frac{1}{\alpha+\frac{1}{2}},\frac{1}{\frac{3}{2}-\beta}\right)<2,$$

because both fractions vary in the open interval (1, 2).

The left-hand side of (33) is, however, a function of class

$$q>\frac{1}{1-\gamma}>2,$$

---

† Remember that the class $(1/\gamma)-0$ of the term $C\phi_\lambda(x)$ is 'higher' than the class q^* of $F(x)$ because $1/q^*>\gamma$.

‡ Remember that the exponents $-(\alpha+\frac{1}{2})$ and $-(\frac{3}{2}-\beta)$ cannot be replaced by higher ones.

because both terms with $\phi_0(x)$ are of class $q^* \geqslant q$ and the term $f(x)$ is of class q.

If $K \neq 0$ there is still a contradiction. Hence, in all cases, $K = 0$, and we see that the function $\phi_0(x)$, and consequently also the function $F(x)$, *actually satisfy the given equation* (3).

We have thus proved:

If (i) *the coefficient-function* $a(x)$ *is continuous in the basic interval* $(-1, 1)$ *and, furthermore, satisfies a Lipschitz condition of positive order* ϵ *in both the arbitrarily small intervals* $(-1, -1+\delta)$, $(1-\delta, 1)$ *with* $\delta > 0$, *and* (ii) *the given function* $f(x)$ *belongs to the class* L_q *with*

$$q > \frac{1}{1-\gamma}, \quad \gamma = \max(\alpha, 1-\beta), \tag{34}$$

then the singular equation (3) *is satisfied by the function*

$$\phi(x) = \frac{a(x)\,\phi(x)}{a^2(x)+\lambda^2\pi^2} + \lambda\pi\frac{A(x)}{1-x}\mathscr{T}_x[(1-y)\,A^*(y)f(y)] + C\frac{A(x)}{1-x} \tag{35}$$

(where C is an arbitrary constant) which belongs at least to a class L_{q^*} *with*

$$q^* < \frac{q}{1+\gamma q}. \tag{36}$$

Moreover, functions of the type

$$C\frac{A(x)}{1-x}$$

are the only non-trivial solutions of the corresponding homogeneous equation in the space L_p $(p > 1)$.

If $q > 1/(1-\beta)$, then the previous solution (35) *can also be put into Carleman's form* (30)–(30').

In particular, if we assume that $a(x) \equiv 0$, which implies that $\theta(x) = \frac{1}{2}\pi$, $\alpha = \beta = \gamma = \frac{1}{2}$, we find again the results of the previous section. The sole difference is that now we have the condition $q > 2$, while there we had the weaker condition $q > \frac{4}{3}$, because of the more accurate determination of the class of $F(x)$.

Recently Carleman's method was used by S. G. Mihlin[†] to solve a singular integral equation which I had encountered (and solved

† 'On the integral equation of F. Tricomi', *Dokl. Akad. Nauk USSR* (N.S.), **59**, 1948, pp. 1053–6.

in another, less simple manner) a long time ago in my research on partial differential equations of mixed type.

Other Russian mathematicians (e.g. I. Vekua, N. I. Muskhelishvili and V. O. Kupradze) have used Carleman's method, particularly for equations with a principal integral over a closed curve in the complex plane. These investigations are contained in the book by Muskhelishvili [32] on this subject, now translated into English, as well as in papers which are not readily available.†

4·5. General Remarks about Non-Linear Integral Equations

Non-linear integral equations are so varied that even classifying them is difficult. Two types, however, are of foremost importance: (i) equations in which either the integrand or the part outside of the integral sign or both depend non-linearly on the unknown function $\phi(x)$, i.e. equations of the form

$$\int_0^1 H[x, y; \phi(y)]\, dy = g[x, \phi(x)];$$ (1)

(ii) equations which contain a truly *non-linear functional*, for instance, the (second) 'integral power'

$$\int_0^1 \int_0^1 K(x, y, z)\, \phi(y)\, \phi(z)\, dy\, dz.$$ (2)

In spite of the fact that one of the most important papers in this field‡ deals with the second type, I believe that the more interesting type is the first, in particular, the case in which the function H is the product of a function K of x and y alone, and a function L of y and $\phi(y)$ alone, i.e. when the equation is of the form

$$\int_0^1 K(x, y)\, L[y, \phi(y)]\, dy = g[x, \phi(x)].$$ (3)

Equations of this kind are discussed in a beautiful paper by Hammerstein.§ Even better, if we can solve the equation

$$g[x, \phi(x)] = \psi(x)$$

---

† See especially, *Trav. Inst. Math. Tbilissi*, **10**, 1942.

‡ E. Schmidt, 'Über die Auflösung der nichtlinearen Integralgleichungen und die Verzweigung ihrer Lösungen', *Math. Ann.* **65**, 1908, pp. 370–99.

§ A. Hammerstein, 'Nichtlineare Integralgleichungen nebst Anwendungen', *Acta Math.* **54**, 1930, 'Ivar Fredholm in memoriam', pp. 117–76.

for $\phi(x)$, then we can consider

$$\psi(x) + \int_0^1 K(x,y)f[y,\psi(y)]\,dy = 0 \qquad (4)$$

as the standard form of *equations of the Hammerstein type*.

This transformation shows that the distinction between homogeneous and non-homogeneous equations, which was so important for linear equations, is almost without significance here. However, it is important to know whether or not the function $f(y,u)$ vanishes identically at $u = 0$ because only in the former case does (4) have the 'trivial' solution $\psi(x) \equiv 0$.

We can use the method of successive approximations to solve (4). For instance, we can put $\psi_0(x) \equiv 0$, and successively

$$\psi_{n+1}(x) = -\int_0^1 K(x,y)f[y,\psi_n(y)]\,dy \qquad (n=0,1,2,\ldots). \qquad (5)$$

However, the convergence can be assured *a priori* only under quite strong conditions on the functions K and f. To be precise, we have to assume:

(i) the kernel K belongs to the class L_2, or, at least, that the function

$$A^2(x) = \int_0^1 K^2(x,y)\,dy \qquad (6)$$

exists almost everywhere in $(0, 1)$ and is summable there;†

(ii) the function $f(y,u)$ satisfies uniformly a Lipschitz condition of the type

$$|f(y,u_1) - f(y,u_2)| < C(y)\,|u_1 - u_2|; \qquad (7)$$

(iii) the function $f(y,0)$ belongs to the class L_2. Under these hypotheses the sequence

$$\psi_0(x), \quad \psi_1(x), \quad \psi_2(x), \quad \ldots \qquad (8)$$

converges almost everywhere to a solution of (4) *provided*

$$\int_0^1 A^2(x)\,C^2(x)\,dx = M^2 < 1. \qquad (9)$$

Moreover, this convergence, is *almost uniform* in the sense of § 2.1.

---

† This is less than $K \in L_2$ because, if $K(x,y)$ belongs to the class L_2, a similar condition must also be satisfied if x and y are interchanged.

In fact, from (5) it follows that

$$| \psi_{n+1}(x) - \psi_n(x) | \leqslant \int_0^1 | K(x,y) | \, |f[y, \psi_n(y)] - f[y, \psi_{n-1}(y)] | \, dy$$

$$< \int_0^1 | K(x,y) | \, C(y) | \, \psi_n(y) - \psi_{n-1}(y) | \, dy,$$

and using the Schwarz inequality, we have

$$| \psi_{n+1}(x) - \psi_n(x) |^2 < \int_0^1 K^2(x,y) \, dy \int_0^1 C^2(y) \, [\psi_n(y) - \psi_{n-1}(y)]^2 \, dy$$

$$= A^2(x) \int_0^1 C^2(y) \, [\psi_n(y) - \psi_{n-1}(y)]^2 \, dy$$

$$(n = 1, 2, 3, \ldots),$$

while in the case $n = 0$, we have

$$\psi_1^2(x) \leqslant A^2(x) \int_0^1 f^2(y, 0) \, dy,$$

or

$$\psi_1^2(x) \leqslant c^2 A^2(x),$$

where c denotes the norm of $f(y, 0)$. Hence, we obtain successively

$$[\psi_2(x) - \psi_1(x)]^2 < c^2 A^2(x) \int_0^1 A^2(y) \, C^2(y) \, dy = c^2 M^2 A^2(x),$$

$$[\psi_3(x) - \psi_2(x)]^2 < c^2 M^2 A^2(x) \int_0^1 A^2(y) \, C^2(y) \, dy = c^2 M^4 A^2(x),$$

$$\ldots\ldots\ldots\ldots\ldots\ldots\ldots\ldots\ldots\ldots\ldots\ldots\ldots\ldots\ldots\ldots\ldots\ldots\ldots\ldots\ldots,$$

and, in general,

$$[\psi_{n+1}(x) - \psi_n(x)]^2 < c^2 M^{2n} A^2(x) \quad (n = 1, 2, 3, \ldots).$$

This shows that the series

$$\psi_1(x) + [\psi_2(x) - \psi_1(x)] + [\psi_3(x) - \psi_2(x)] + \ldots \tag{10}$$

has the majorant $\quad cA(x) \, (1 + M + M^2 + \ldots),$

which converges for $M < 1$; hence the sequence (8) converges almost uniformly under the stated conditions.

Furthermore, it is easy to show, in the usual way, that the sum of series (10), i.e. the limit of $\psi_n(x)$ as $n \to \infty$, is a solution of the given equation (4).

The importance of the previous conditions and especially of inequality (9) need not be pointed out, since there are very simple non-linear integral equations which have *no solutions* at all. For instance, if

$$K(x, y) = \alpha(x)\,\alpha(y), \tag{11}$$

so that necessarily

$$\psi(x) = \xi\alpha(x) \quad (\xi = \text{constant}),$$

then equation (4) becomes equivalent to the non-linear equation for ξ

$$\xi + \int_0^1 \alpha(y)f[y, \xi\alpha(y)]\,dy = 0. \tag{12}$$

It is very easy to construct examples for which no real solutions of this equation exist. For instance, if

$$f(y, u) = \tfrac{1}{2}(1 + u^2),$$

equation (12) becomes

$$\xi^2\int_0^1 \alpha^3(y)\,dy + 2\xi + \int_0^1 \alpha(y)\,dy = 0.$$

This equation of the second degree has no real solutions if

$$\int_0^1 \alpha^2(y)\,dy > 1, \tag{13}$$

since, by the Schwarz inequality,

$$1 - \int_0^1 \alpha^3(y)\,dy\int_0^1 \alpha(y)\,dy \leqslant 1 - \left[\int_0^1 \alpha^2(y)\,dy\right]^2.$$

On the other hand, if $\alpha(y)$ is always positive, and we set

$$f(y, u) = u\sin\frac{u}{\alpha(y)},$$

then equation (12) becomes

$$1 + \sin\xi\int_0^1 \alpha^2(y)\,dy = 0,$$

which has an *infinity of real solutions* provided condition (13) is satisfied.

Finally, we shall give a *necessary* condition for a point $\lambda = \lambda_0$ to be a bifurcation point *for a given solution* $\psi_0(x)$ of a non-linear equation with a parameter λ, of the type

$$\psi(x) + \lambda\int_0^1 K(x, y)f[y, \psi(y)]\,dy = 0, \tag{14}$$

i.e. a condition for the presence, on the left or on the right of the point λ_0 on the λ-axis, of (at least) another solution $\psi_1(x)$ of the equation, such that the quotient

$$\frac{\psi_1(x) - \psi_0(x)}{\lambda - \lambda_0}$$

approaches a limit function $\chi_0(x)$, as $\lambda \to \lambda_0$, which does not vanish almost everywhere. If the derivative f_u' of $f(y, u)$ with respect to u exists and is continuous, then this condition is that λ_0 *be an eigenvalue of the Fredholm kernel*

$$K^*(x, y) = -K(x, y) f_u'[y, \psi_0(x)]. \tag{15}$$

To show this, set

$$\lambda - \lambda_0 = \epsilon, \quad \psi_1(x) - \psi_0(x) = \epsilon \chi(x).$$

Then, since $\psi_1(x)$ is a solution of (14), we have identically

$$\psi_0(x) + \epsilon \chi(x)$$

$$+ \lambda \int_0^1 K(x, y) \{f[y, \psi_0(y) + \epsilon \chi(y)] - f[y, \psi_0(y)] + f[y, \psi_0(y)]\} \, dy = 0.$$

On the other hand,

$$\psi_0(x) + \lambda \int_0^1 K(x, y) f[y, \psi_0(y)] \, dy = 0;$$

hence, subtracting and then dividing by ϵ we have

$$\chi(x) + \lambda \int_0^1 K(x, y) \frac{f[y, \psi_0(x) + \epsilon \chi(y)] - f[y, \psi_0(y)]}{\epsilon \chi(y)} \chi(y) \, dy = 0.$$

Letting $\lambda \to \lambda_0$, $\epsilon \to 0$, we obtain

$$\chi_0(x) - \lambda_0 \int_0^1 K^*(x, y) \chi_0(y) \, dy = 0, \tag{16}$$

where $\chi_0(x)$ is the limit of $\chi(x)$ as $\epsilon \to 0$. Hence $\chi_0(x)$ must be a nontrivial solution of the previous homogeneous linear integral equation with kernel K^*.†

† Notice that the previous condition must be satisfied also in the case in which the neighboring solution $\psi_1(x)$ exists on the left as well as on the right of the point $\lambda = \lambda_0$. Then, according to our definition (p. 161), λ_0 is no longer a bifurcation point. Hence the previous condition is more a condition for the 'meeting' of two or more solutions of the equation than a condition for the existence of a bifurcation point.

4·6. Non-Linear Equations of the Hammerstein Type

In a paper mentioned in the previous section, A. Hammerstein studied non-linear integral equations of the type†

$$\psi(x) + \int_0^1 K(x,y) f[y, \psi(y)] \, dy = 0, \qquad (1)$$

using the following basic hypotheses:

(i) the fundamental theorem of Fredholm is valid for the linear integral equation with the kernel K,‡ and the iterated kernel $K_2(x,y)$ is continuous;

(ii) the kernel K is symmetric, i.e. $K(x,y) = K(y,x)$;

(iii) the kernel K is *positive*, i.e. all its eigenvalues are positive.

If these conditions are satisfied, we shall say that the integral equation belongs *properly* to Hammerstein's type.

Hammerstein's method is centered about the Hilbert-Schmidt theorem. To be precise, he uses the fact that since by (1)

$$\psi(x) = \int_0^1 K(x,y) g(y) \, dy \quad \text{with} \quad g(y) = -f[y, \psi(y)],$$

if it exists at all and $g(y) \in L^2$, $\psi(x)$ can be represented by a (uniformly) convergent series of the form

$$\psi(x) = \sum_{m=1}^{\infty} c_m \phi_m(x), \qquad (2)$$

where $\phi_1(x)$, $\phi_2(x)$, ... are the (orthonormalized) eigenfunctions of the kernel $K(x,y)$ (corresponding to the eigenvalues λ_1, λ_2, ..., respectively), and c_1, c_2, ... are certain unknown constants.

---

† Hammerstein actually deals with integrals extending over n-dimensional manifolds (like equation (6) of § 3·17); however, this implies no important differences in the treatment.

‡ A sufficient condition for this is for K to belong to the class L_2, in the sense of Chapter II. Moreover, for the sake of simplicity, we shall now suppose even that $K_2(x,y)$ *is continuous*, which, in view of the symmetry of K, implies $K \in L_2^*$, since

$$A^2(x) \equiv B^2(x) \equiv \int_0^1 K^2(x,y) \, dy = K_2(x,x).$$

Consequently the series of the Hilbert-Schmidt theorem will not only be *almost uniformly* convergent, but even *uniformly* convergent. The hypothesis that K_2 be continuous is implicit in the paper of Hammerstein.

Therefore, since

$$c_m = \int_0^1 \psi(x)\,\phi_m(x)\,dx = -\int_0^1 \phi_m(x)\,dx \int_0^1 K(x,y)f[y,\psi(y)]\,dy$$

$$= -\int_0^1 f[y,\psi(y)]\,dy \int_0^1 K(x,y)\phi_m(x)\,dx = -\frac{1}{\lambda_m}\int_0^1 f[y,\psi(y)]\,\phi_m(y)\,dy,$$

the problem of solving the given equation is equivalent to that of solving the following system of infinite equations with an infinite number of unknowns:

$$c_m = -\frac{1}{\lambda_m}\int_0^1 f\left[y,\sum_{h=1}^{\infty} c_h\phi_h(y)\right]\phi_m(y)\,dy \quad (m=1,2,3,\ldots). \quad (3)$$

It is now quite natural to consider the 'approximate' solution

$$\psi_n(x) = \sum_{m=1}^{n} c_{n,m}\phi_m(x), \quad (4)$$

where the n constants $c_{n,1}, c_{n,2}, \ldots, c_{n,n}$ have to satisfy the system of n equations with n unknowns

$$c_{n,m} = -\frac{1}{\lambda_m}\int_0^1 f\left[y,\sum_{h=1}^{n} c_{n,h}\phi_h(y)\right]\phi_m(y)\,dy \quad (m=1,2,\ldots,n). \quad (5)$$

But, does this system have any solution at all?

By considering a minimum problem for a suitable function of n variables, Hammerstein showed very elegantly that *the system* (5) *has at least one solution, provided that the function $f(x,u)$ is continuous and its absolute value is less than a suitable linear function of* $|u|$, *i.e. that it satisfies a condition of the type*

$$|f(x,u)| \leqslant C_1|u| + C_2, \quad (6)$$

where C_1 and C_2 are two positive constants and C_1 is less than the first eigenvalue λ_1 of the positive kernel $K(x,y)$.†

To prove this, Hammerstein considered the continuous function

$$H(x_1, x_2, \ldots, x_n) = \sum_{m=1}^{n} \lambda_m x_m^2 + 2\int_0^1 F\left[y,\sum_{h=1}^{n} x_h\phi_h(y)\right]dy,$$

where

$$F(y,u) = \int_0^u f(y,v)\,dv. \quad (7)$$

---

† Despite the fact that inequality (6) can be weakened somewhat (see below), Hammerstein showed that the limitation $C_1 < \lambda_1$ cannot (in general) be improved.

$H(x_1, x_2, ..., x_n)$ is a function whose partial derivatives are closely related to the equations of system (5) since

$$\frac{1}{2\lambda_m}\frac{\partial H}{\partial x_m} = x_m + \frac{1}{\lambda_m}\int_0^1 f\left[y, \sum_{h=1}^n x_h\phi_h(y)\right]\phi_m(y)\,dy. \tag{8}$$

It is not difficult to see that the function H is *bounded from below*. In fact, from (6) and (7) it follows that

$$|F(x, u)| \leqslant \tfrac{1}{2}C_1 u^2 + C_2|u|, \tag{9}$$

and, if C_1 is less than an arbitrary constant k, in view of the elementary inequality†

$$ax - bx^2 \leqslant \frac{a^2}{4b} \quad (b > 0), \tag{10}$$

by identifying x with $|u|$, a with C_2 and b with $\tfrac{1}{2}(k - C_1)$, we have

$$\tfrac{1}{2}C_1 u^2 + C_2|u| \leqslant \tfrac{1}{2}ku^2 + C_3,$$

where

$$C_3 = \frac{C_2^2}{2(k - C_1)}.$$

Hence if

$$C_1 < k < \lambda_1,$$

we have

$$F(x, u) \geqslant -(\tfrac{1}{2}ku^2 + C_3). \tag{11}$$

Consequently, we obtain

$$H(x_1, x_2, ..., x_n) \geqslant \sum_{m=1}^n \lambda_m x_m^2 - \int_0^1\left[k\left(\sum_{h=1}^n x_h\phi_h(y)\right)^2 + 2C_3\right]dy,$$

that is,

$$H \geqslant \sum_{m=1}^n \lambda_m x_m^2 - k\sum_{h=1}^n x_h^2 - 2C_3 = \sum_{m=1}^n (\lambda_m - k)x_m^2 - 2C_3. \tag{12}$$

Since $k < \lambda_1 \leqslant \lambda_2 \leqslant \lambda_3 \leqslant ...$, the sum on the right is not negative. It follows that H has the lower bound $-2C_3$.

Hence there is at least one set $x_1^{(0)}, x_2^{(0)}, ..., x_n^{(0)}$ of values $x_1, x_2, ..., x_n$ for which the continuous function H becomes an *absolute minimum* d_n. We can be certain that by putting

$$c_{n,m} = x_m^{(0)} \quad (m = 1, 2, ..., n)$$

† Multiplying by the positive number $4b$, (10) is equivalent to the statement that

$$a^2 - 4abx + 4b^2x^2 = (a - 2bx)^2 \geqslant 0.$$

the system (5) *will be satisfied*, because for

$$x_m = x_m^{(0)} = c_{n,m} \quad (m = 1, 2, \ldots, n), \tag{13}$$

we must have

$$\frac{\partial H}{\partial x_1} = \frac{\partial H}{\partial x_2} = \ldots = \frac{\partial H}{\partial x_n} = 0.$$

Furthermore, it is important to show that, with this choice of the quantities $c_{n,m}$, the sum

$$S_n = \sum_{m=1}^{n} \lambda_m c_{n,m}^2 \tag{14}$$

has an upper bound independent of n.

In fact, when the x_m have the values (13), it follows from (12) that

$$\sum_{m=1}^{n} (\lambda_m - k) c_{n,m}^2 \leqslant d_n + 2C_3,$$

and *a fortiori*

$$\left(1 - \frac{k}{\lambda_1}\right) \sum_{m=1}^{n} \lambda_m c_{n,m}^2 = \sum_{m=1}^{n} \left(\lambda_m - k\frac{\lambda_m}{\lambda_1}\right) c_{n,m}^2 \leqslant \sum_{m=1}^{n} (\lambda_m - k) c_{n,m}^2 \leqslant d_n + 2C_3,$$

that is,

$$S_n \leqslant \frac{d_n + 2C_3}{1 - k/\lambda_1}.$$

But, on the other hand, since

$$H_n(x_1, x_2, \ldots, x_n) = H_{n+1}(x_1, x_2, \ldots, x_n, 0),$$

we have

$$d_{n+1} \leqslant d_n \quad (n = 1, 2, 3, \ldots).$$

Hence $d_n \leqslant d_1$ and we obtain

$$S_n \leqslant D = \frac{d_1 + 2C_3}{1 - k/\lambda_1}. \tag{15}$$

This shows that we can also write

$$\int_0^1 \psi_n^2(x)\,dx = \sum_{m=1}^{n} c_{n,m}^2 \leqslant \frac{1}{\lambda_1} \sum_{m=1}^{n} \lambda_m c_{n,m}^2 \leqslant \frac{D}{\lambda_1}. \tag{16}$$

Now we have to prove that as $n \to \infty$ the function $\psi_n(x)$ approaches a solution of the given equation.

We shall show first that the function

$$\chi_n(x) = \psi_n(x) + \int_0^1 K(x,y) f[y, \psi_n(y)]\,dy \tag{17}$$

approaches zero uniformly, as $n \to \infty$.

In fact, by using the Hilbert-Schmidt theorem again, we obtain

$$\chi_n(x) = \sum_{m=1}^{n} c_{n,m} \phi_m(x) + \sum_{m=1}^{\infty} \phi_m(x) \int_0^1 \phi_m(\xi)\, d\xi \int_0^1 K(\xi, y) f[y, \psi_n(y)]\, dy$$

$$= \sum_{m=1}^{n} c_{n,m} \phi_m(x) + \sum_{m=1}^{\infty} \frac{\phi_m(x)}{\lambda_m} \int_0^1 f[y, \psi_n(y)] \phi_m(y)\, dy.$$

By virtue of (5), we have

$$\frac{1}{\lambda_m} \int_0^1 f[y, \psi_n(y)] \phi_m(y)\, dy = -c_{n,m}.$$

Hence the first n terms of the second sum annihilate the first one, and we obtain

$$\chi_n(x) = \sum_{m=n+1}^{\infty} \lambda_m^{-1} \phi_m(x) \int_0^1 f[y, \psi_n(y)] \phi_m(y)\, dy.$$

Consequently, with the help of the Schwarz inequality for sums, we can state that

$$\chi_n^2(x) \leqslant \sum_{m=n+1}^{\infty} \lambda_m^{-2} \phi_m^2(x) \sum_{m=n+1}^{\infty} \left[\int_0^1 f[y, \psi_n(y)] \phi_m(y)\, dy \right]^2. \quad (18)$$

Now, using the Schwarz inequality for integrals, we have

$$\left\{ \int_0^1 f[y, \psi_n(y)] \phi_m(y)\, dy \right\}^2 \leqslant \int_0^1 \phi_m^2(y)\, dy \int_0^1 f^2[y, \psi_n(y)]\, dy$$

$$= \int_0^1 f^2[y, \psi_n(y)]\, dy.$$

Squaring both sides of (6) gives

$$f^2(x, u) \leqslant C_1^2 u^2 + 2C_1 C_2 \mid u \mid + C_2^2 = 2C_1 C_2 \mid u \mid - (k - C_1^2) u^2 + ku^2 + C_2^2.$$

Therefore, if we assume that $C_1^2 < k$ and apply inequality (10) with $x = \mid u \mid$, $a = 2C_1 C_2$, $b = k - C_1^2$, we get

$$f^2(y, u) \leqslant ku^2 + C_4,$$

where C_4 is a suitable constant. Hence, remembering (16), we obtain

$$\int_0^1 f^2[y, \psi_n(y)]\, dy \leqslant \int_0^1 [k\psi_n^2(y) + C_4]\, dy \leqslant \frac{k}{\lambda_1} D + C_4 = D^*, \quad (19)$$

and inequality (18) becomes

$$\chi_n^2(x) \leqslant D^* \sum_{m=n+1}^{\infty} \lambda_m^{-2} \phi_m^2(x).$$

It is well known that the infinite series $\Sigma \lambda_m^{-2} \phi_m^2(x)$ converges uniformly to $K_2(x, x)$. Hence $\chi_n(x)$ converges uniformly to *zero* as $n \to \infty$.

In this way we have proved that *if the sequence* $\psi_1(x)$, $\psi_2(x)$, ... *converges to a limit function* $\psi(x)$, and if Lebesgue's fundamental integral theorem can be applied to the evaluation of the limit as $n \to \infty$ of the integral

$$\int_0^1 K(x,y) f[y, \psi_n(y)] \, dy,$$

then the limit function $\psi(x)$ is a solution of the given equation (1). But does the sequence ψ_1, ψ_2, ... converge at all? Under the previous hypotheses we cannot assert that it does, but we can easily show that from the sequence $\psi_1(x)$, $\psi_2(x)$, ... *there can always be extracted a subsequence*

$$\psi_{n_1}(x), \ \psi_{n_2}(x), \psi_{n_3}(x), \ldots$$

which converges uniformly to a (continuous) limit function $\psi(x)$.

To do this, it is sufficient to apply to the sequence

$$\omega_n(x) = \chi_n(x) - \psi_n(x) = \int_0^1 K(x,y) f[y, \psi_n(y)] \, dy \quad (n = 1, 2, 3, \ldots)$$

a well-known theorem,† which permits us to deduce the previous assertion from the uniform boundedness and equicontinuity of the considered functions.

The sequence $\{\omega_n\}$ is uniformly bounded because, as a consequence of (19), we have

$$\omega_n^2(x) \leqslant \int_0^1 K^2(x,y) \, dy \int_0^1 f^2[y, \psi_n(y)] \, dy \leqslant D^* \int_0^1 K^2(x,y) \, dy.$$

The functions $\omega_n(x)$ are *equicontinuous* in view of the inequalities

$$[\omega_n(x_1) - \omega_n(x_2)]^2 = \left\{ \int_0^1 [K(x_1,y) - K(x_2,y)] f[y, \psi_n(y)] \, dy \right\}^2$$

$$\leqslant \int_0^1 [K(x_1,y) - K(x_2,y)]^2 \, dy \int_0^1 f^2[y, \psi_n(y)] \, dy$$

$$\leqslant D^* \int_0^1 [K^2(x_1,y) - 2K(x_1,y)K(x_2,y) + K^2(x_2,y)] \, dy$$

$$= D^*[K_2(x_1,x_1) - 2K_2(x_1,x_2) + K_2(x_2,x_2)]$$

$$= D^*[K_2(x_1,x_1) - K_2(x_1,x_2)] + D^*[K_2(x_2,x_2) - K_2(x_1,x_2)],$$

† See, for example, Courant-Hilbert [6], p. 39 or Goffman (op. cit., in § 2·1), p. 106, Th. 5.

and of the fact that the iterated kernel $K_2(x, y)$ is assumed to be continuous.

Since $\lim \chi_n(x) = 0$, the passage from the sequence $\{\omega_n\}$ to the sequence $\{\psi_n\}$ is obvious.

Therefore, we have proved the following:

EXISTENCE THEOREM. *If the kernel K satisfies the basic conditions* (i), (ii), (iii), *and the continuous function $f(y, u)$ satisfies condition* (6), *then the non-linear integral equation* (1) *has at least one (continuous) solution.*

Condition (6) may be weakened in different ways. For instance, it is quite obvious that, *if we keep condition* (10) *on $F(y, u)$*, then the continuous function $f(y, u)$ need only be such that

$$|f(y, u)| < C_1 |u| + C_2, \tag{6'}$$

where C_1 and C_2 are *any* positive constants, C_1 not necessarily less than λ_1.

Furthermore, if the kernel K is *continuous*, hence bounded, we can dispose of condition (6) completely, and keep only (11), because an inequality of the type (19) can now be deduced, independently of (6), from the fact that the function $f(y, u)$ is continuous and $\psi_n(y)$ is *bounded*. In fact, we can write

$$\psi_n(x) = \sum_{m=1}^{n} c_{n, m} \phi_m(x) = \sum_{m=1}^{\infty} \sqrt{\lambda_m} c_{n, m} \frac{\phi_m(x)}{\sqrt{\lambda_m}},$$

and consequently, in view of Mercer's theorem, we obtain

$$\psi_n^2(x) \leqslant S_n \sum_{m=1}^{n} \frac{\phi_m^2(x)}{\lambda_m} \leqslant S_n \sum_{m=1}^{\infty} \frac{\phi_m^2(x)}{\lambda_m} = S_n K(x, x) \leqslant D \max K(x, x). \tag{20}$$

Another important point is that *if the values of the kernel $K(x, y)$ and of the function $f(y, u)$ are never negative*,

$$K(x, y) \geqslant 0, \quad f(y, u) \geqslant 0, \tag{21}$$

then the validity of the existence theorem can be assured by imposing suitable restrictions on the continuous function $f(y, u)$ for $u \leqslant 0$ only, e.g. by supposing that

$$0 \leqslant f(y, u) \leqslant C_1 |u| + C_2 \quad (u \leqslant 0, \ 0 < C_1 < \lambda_1). \tag{22}$$

In fact, if we put

$$*(y, u) = \begin{cases} f(y, u) & (u \leqslant 0), \\ f(y, 0) & (u \geqslant 0), \end{cases}$$

then the continuous function $f^*(y, u)$ satisfies condition (6), and consequently the equation

$$\psi^*(x) = -\int_0^1 K(x, y) f^*[y, \psi^*(y)] \, dy \tag{23}$$

has at least one solution Ψ^*, which can be determined as we have shown above. But from the hypotheses $K \geqslant 0, f^* \geqslant 0$, it follows that $\psi^* \leqslant 0$; hence

$$f^*[y, \psi^*(y)] = f[y, \psi^*(y)].$$

This shows that the previous equation differs from (1) in appearance only; consequently this equation has at least the solution Ψ^*.

The validity of the previous observation can be extended by first noting that, if we set $\psi(x) = -\psi_1(x)$, equation (1) is carried into the similar one

$$\psi_1(x) + \int_0^1 K(x, y) f_1[y, \psi_1(y)] \, dy = 0, \tag{24}$$

where

$$f_1(y, u) = -f(y, -u). \tag{25}$$

Secondly, we observe that by putting

$$f(y, u) = f_1(y, u) - C_3,$$

where C_3 is any constant (but we will assume $C_3 > 0$), the given equation becomes

$$\psi(x) - k(x) + \int_0^1 K(x, y) f_1[y, \psi(y)] \, dy = 0,$$

where

$$k(x) = C_3 \int_0^1 K(x, y) \, dy.$$

Consequently, if for $\psi(x)$ we substitute

$$\psi^*(x) = \psi(x) - k(x),$$

and if we set

$$f^*(y, u) = f_1[y, u + k(y)] = f[y, u + k(y)] + C_3,$$

then the given equation remains unchanged, in the sense that it becomes

$$\psi^*(x) + \int_0^1 K(x, y) f^*[y, \psi^*(y)] \, dy = 0.$$

Note that to the condition

$$0 \leqslant f^*(y, u) \leqslant C_1 |u| + C_2 \quad (u \leqslant 0)$$

there corresponds the condition

$$-C_3 \leqslant f[y, u+k(y)] \leqslant C_1 |u| + C_2 - C_3 \quad (u \leqslant 0),$$

and note further that by setting

$$u + k(y) = v,$$

since $k(y) > 0$, we have

$$v > u, \quad |v| < |u| \quad (v \leqslant 0).$$

We see thus that the existence theorem holds also when (with $C_1 < \lambda_1$ and C_4 an arbitrary constant)

$$\left.\begin{array}{ll} -C_3 \leqslant f(y,v) \leqslant C_1 |v| + C_4 & (v \leqslant 0), \\ \quad f(y,v) \geqslant -C_3 & (v \geqslant 0), \end{array}\right\} \tag{26}$$

or, using (25), when

$$f(y,v) \leqslant C_3 \quad (v \leqslant 0), \quad -(C_1 v + C_2) \leqslant f(y,v) \leqslant C_3 \quad (v \geqslant 0). \tag{27}$$

All this allows us to establish the following table of the principal cases in which the existence of at least one (continuous) solution of the non-linear equation (1) can be assured *a priori*:

Cases	Hypo-thesis† about $K(x,y)$	Hypothesis‡ about $f(y,u)$ if $u \leqslant 0$	if $u \geqslant 0$	Examples for $f(y,u)$
a	—	f bounded or satisfying (6)		$\sin u$
b	Continuous	F bounded *below* or satisfying (11)		e^u
c		$0 \leqslant f \leqslant C_1 |u| + C_2$	$f \geqslant 0$	$1 + e^u \sin^2 y$
d		$f \leqslant 0$	$-(C_1 u + C_2) \leqslant f \leqslant 0$	$-(1 + e^{-u} \times \sin^2 y)$
e	$K(x,y) \geqslant 0$	$-C_3 \leqslant f \leqslant C_1 |u| + C_2$	$f \geqslant -C_3$	$p(y) e^u + q(y)$ ($p \geqslant 0$, p, q cont.)
f		$f \leqslant C_3$	$-(C_1 u + C_2) \leqslant f \leqslant C_3$	$-p(y) e^{-u} + q(y)$ ($p \geqslant 0$; p, q cont.)

† In addition to the basic hypotheses (i), (ii) and (iii).

‡ In addition to continuity. Moreover, it is supposed that $0 < C_1 < \lambda_1$.

In Hammerstein's paper there are also some interesting remarks about the uniqueness or non-uniqueness of the solutions of (1). They begin with the observation that the existence of two different solutions $\psi(x)$, $\psi^*(x)$ of (1) implies the existence of a non-trivial solution of the equation

$$\Psi(x) + \int_0^1 K(x, y)\, g[y, \Psi(y)]\, dy = 0, \qquad (28)$$

where

$$\Psi(x) = \psi^*(x) - \psi(x), \quad g(y, u) = f[y, u + \psi(y)] - f[y, \psi(y)], \qquad (29)$$

and vice versa.

A simple but interesting result concerning uniqueness is the observation that *if*

$$g(y, u) \geqslant 0 \quad (u > 0), \qquad g(y, u) \leqslant 0 \quad (u < 0), \qquad (30)$$

then equation (28) *has only the solution* $\Psi(x) \equiv 0$.

In fact, from (28) it follows that

$$\int_0^1 \Psi(x)\, g[x, \Psi(x)]\, dx = - \int_0^1 \int_0^1 K(x, y)\, g[x, \Psi(x)]\, g[y, \Psi(y)]\, dx\, dy.$$

Hypothesis (30) shows that

$$\int_0^1 \Psi(x)\, g[x, \Psi(x)]\, dx \geqslant 0,$$

and the positive character of the kernel K shows that

$$-\int_0^1 \int_0^1 K(x, y)\, g[x, \Psi(x)]\, g[y, \Psi(y)]\, dx\, dy < 0,$$

provided we do not have

$$g[x, \Psi(x)] \equiv 0. \qquad (31)$$

Hence condition (31) must necessarily be satisfied and then from (28) it follows that $\Psi(x) \equiv 0$.

Since the function $g(y, u)$, given by the second equality (29), obviously satisfies condition (30) when $f(y, u)$ is a non-decreasing monotone function of u, we obtain the following theorem:

UNIQUENESS THEOREM I. *If for any fixed y in* $(0, 1)$, *the function* $f(y, u)$ *is a non-decreasing function of* u, *then the non-linear integral equation* (1) *has at most one solution.*

An equally important theorem is

UNIQUENESS THEOREM II. *The non-linear integral equation* (1) *has at most one solution if the function* $g(y, u)$ *satisfies an inequality of the type*

$$| g(y, u) | \leqslant \alpha | u | \quad (0 < \alpha < \lambda_1). \tag{32}$$

For the proof we notice that, in view of the Hilbert-Schmidt Theorem, we can write

$$\Psi(x) = \sum_{m=1}^{\infty} \gamma_m \phi_m(x),$$

where

$$\gamma_m = \int_0^1 \Psi(x) \phi_m(x) \, dx = -\int_0^1 \phi_m(x) \, dx \int_0^1 K(x, y) g[y, \Psi(y)] \, dy$$

$$= -\frac{1}{\lambda_m} \int_0^1 g[y, \Psi(y)] \phi_m(y) \, dy.$$

Consequently the product $-\lambda_m \gamma_m$ is the mth Fourier coefficient of the function $g[y, \Psi(y)]$ and, by virtue of Bessel's inequality and (32), we obtain

$$\sum_{m=1}^{\infty} \lambda_m^2 \gamma_m^2 \leqslant \int_0^1 g^2[y, \Psi(y)] \, dy \leqslant \alpha^2 \int_0^1 \Psi^2(y) \, dy = \alpha^2 \sum_{m=1}^{\infty} \gamma_m^2,$$

that is,

$$\sum_{m=1}^{\infty} (\alpha^2 - \lambda_m^2) \gamma_m^2 \geqslant 0.$$

But, since

$$\alpha^2 - \lambda_m^2 \leqslant \alpha^2 - \lambda_1^2 < 0,$$

the previous inequality is an absurdity unless

$$\gamma_1 = \gamma_2 = \ldots = 0,$$

hence $\Psi(x) \equiv 0$.

From the previous theorem, the next follows immediately:

UNIQUENESS THEOREM III. *The non-linear integral equation* (1) *has at most one solution provided that the function* $f(y, u)$ *satisfies uniformly a Lipschitz condition of the type*

$$| f(y, u_1) - f(y, u_2) | < \alpha | u_1 - u_2 |, \tag{33}$$

where the positive constant α *is less than the first eigenvalue* λ_1 *of the kernel* K.

In fact, under the previous hypothesis, the function $g(y, u)$ given by (29) obviously satisfies condition (32).

Finally we note that, among others, Golomb,[†] Nazarow,[‡] and Trjitzinsky[§] have studied more general non-linear integral equations than those of Hammerstein's type, especially equations of the form (1) with a *singular* kernel K.

4·7. Forced Oscillations of Finite Amplitude

The usefulness of the results of the previous section may be pointed out by applying them to a celebrated and difficult problem: the study of the forced oscillations of *finite amplitude* of a pendulum, which, in the absence of a damping force, are governed by the non-linear differential equation

$$\frac{d^2\phi(t)}{dt^2} + \alpha^2 \sin \phi(t) = F(t). \tag{1}$$

The main difficulty arises from the fact that, for a given α^2 and a given periodic 'driving function' $F(t)$, we can have either 'resonance' or 'non-resonance' depending on the amplitude of the oscillations.

In fact, while the period of the 'infinitely small' free oscillations of the pendulum is the constant $2\pi/\alpha$ ($\alpha^2 = g/l$), because the oscillations are governed by the linear equation

$$\frac{d^2\phi}{dt^2} + \alpha^2\phi = 0, \tag{2}$$

the period of the free oscillation of finite amplitude ϕ_0 is given by

$$\frac{4}{\alpha}K(k) \quad \text{with} \quad k = \sin(\tfrac{1}{2}\phi_0),$$

where $K(k)$ denotes the complete elliptic integral of Legendre:

$$K(k) = \int_0^{\frac{1}{2}\pi} \frac{d\theta}{\sqrt{(1 - k^2 \sin^2 \theta)}},$$

† M. Golomb, 'Zur Theorie der nichtlinearen Integralgleichungen usw.', *Math. Zeitschr.* **39**, 1935, pp. 45–75.

‡ N. Nazarow, 'Non-linear integral equations of Hammerstein's type', *Acta (Trudi) Univ. Asiae Mediae*, Ser. v, fasc. 28, 1939, 12 pp.; fasc. 33, 1941, pp. 3–78; N.S. fasc. 6, 1945, 14 pp.; *Math. Rev.* **3**, 1942, p. 150; **8**, 1947, p. 518.

§ W. J. Trjitzinsky, 'Singular non-linear integral equations', *Duke Math. J.* **11**, 1944, pp. 517–64.

which (see Fig. 14) varies monotonically from $\frac{1}{2}\pi$ to ∞ as $k = \sin(\frac{1}{2}\phi_0)$ increases from 0 to 1.

After publication of a paper by Duffing,[†] Hamel,[‡] and later Iglisch,[§] studied this problem. They used essentially the methods of Hammerstein and Schmidt.

Suppose that the driving function $F(t)$ is an *odd* function with a certain period 2ω, as happens in the important case

$$F(t) = \beta \sin(\pi t/\omega),$$

then the main question to ask about (1) is whether or not this equation has a solution of the same kind, i.e. a solution which is also odd and periodic with the period 2ω.

In order to solve this difficult problem we first observe that, if $F(t)$ satisfies the stated condition, then from any solution $\phi^*(t)$ of (1) defined only for $0 \leqslant t \leqslant \omega$ and such that

$$\phi^*(0) = \phi^*(\omega) = 0, \tag{3}$$

we can immediately obtain an odd periodic solution of (1) (with the period ω) by setting

$$\phi(\tau + n\omega) = \phi^*(\tau) \quad (0 \leqslant \tau \leqslant \omega, \; n = 0, 1, 2, \ldots)$$

and $$\phi(-t) = -\phi(t).$$

Conversely, any odd periodic solution of (1) obviously must satisfy conditions (3).

We thus see that our problem is equivalent to that of solving the non-linear differential equation (1) together with the classical boundary conditions

$$\phi(0) = \phi(1) = 0, \tag{4}$$

where, for the sake of simplicity, we put $\omega = 1$.

† G. Duffing, *Erzwungene Schwingungen bei veränderlicher Eigenfrequenz und ihre technische Bedeutung*, Sammlung Vieweg, Heft 41/42, Braunschweig, 1918.

‡ G. Hamel, 'Über erzwungene Schwingungen bei endlichen Amplituden', *Math. Ann.* **86**, 1922, 2.

§ See, for example, Tricomi, 'Elliptische Funktionen', Leipzig, *Akad. Verlagsgesell.* 1948, pp. 272–4 (2nd Italian ed. Bologna, Zanichelli, 1951); R. Iglisch, *Monatsh. Math. Physik* (Wien), **37**, 1930, pp. 325–42; **39**, 1932, pp. 173–220; **42**, 1935, pp. 7–36; *Math. Ann.* **111**, 1935, pp. 568–81; **112**, 1936, pp. 221–46.

We now remember that the Green's function of the Sturm-Liouville system

$$y''(x) + \lambda r(x) y(x) = 0, \quad y(0) = y(1) = 0$$

is the *triangular kernel*

$$T(x,y) = \begin{cases} x(1-y) & (0 \leqslant x \leqslant y), \\ y(1-x) & (y \leqslant x \leqslant 1). \end{cases}$$

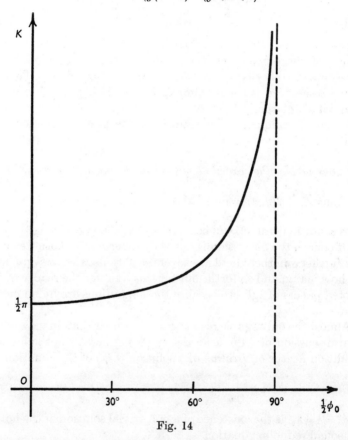

Fig. 14

We remember also that the solution of a non-homogeneous differential equation can be expressed by means of (3·13-30). Consequently system (1) + (4) is equivalent to the non-linear integral equation

$$\phi(x) = -\int_0^1 T(x,y)\,[F(y) - \alpha^2 \sin\phi(y)]\,dy,$$

which, if we put

$$\int_0^1 T(x,y)\,F(y)\,dy = g(x), \quad \phi(x) + g(x) = \psi(x), \qquad (5)$$

assumes Hammerstein's form

$$\psi(x) + \int_0^1 T(x,y)f[y,\psi(y)]\,dy = 0 \qquad (6)$$

with
$$f(y,u) = \alpha^2 \sin[u - g(y)]. \qquad (7)$$

The function $f(y,u)$ is actually *bounded*. Consequently the results of the previous section (case a of the table) allow us to state immediately that *our problem has at least one solution*. This is a trivial result if α^2 is small (compared to 1), but it is certainly not trivial if α^2 is large.†

Moreover, uniqueness Theorem II of the previous section tells us that if
$$\alpha < \pi, \qquad (8)$$

then *our problem has exactly one solution*, because we now have

$$\max \left| \frac{\partial}{\partial u} f(y,u) \right| = \max | \alpha^2 \cos[u - g(y)] | = \alpha^2, \quad \lambda_1 = \pi^2;$$

thus condition (4·6-32) will be certainly satisfied provided (8) holds.

Of course, the discussion is not yet exhausted, because we could ask further whether there are several solutions when $\alpha > \pi$, and if so, how many, and so forth; but for this we refer the reader to the quoted papers of Iglisch, because these questions are by no means simple.

Almost the only simple result of this nature is that, in view of the final discussion of § 4·5, a necessary (but generally not sufficient) condition for the *bifurcation* of a solution $\psi_0(x)$ of the equation

$$\psi(x) + \lambda \int_0^1 T(x,y)\sin[\psi(y) - g(y)]\,dy = 0 \qquad (9)$$

at $\alpha^2 = \lambda = \lambda_0$, is the existence of a non-trivial solution of the homogeneous Fredholm equation

$$\phi(x) - \lambda_0 \int_0^1 T(x,y)\cos[\psi_0(y) - g(y)]\,\phi(y)\,dy = 0. \qquad (10)$$

---

† It may seem strange that we can have 'resonance' even when $2\pi/\alpha$ is very large relative to ω, but there will always be entire multiples of ω near or larger than $2\pi/\alpha$.

In other words, λ_0 must be an eigenvalue of the kernel

$$K(x,y) = T(x,y) \cos[\psi_0(y) - g(y)].$$

Thus we see that the only *possible* bifurcation points of the solutions of our problem for a pendulum *at rest*, i.e. for the solution $\psi_0(x) \equiv g(x)$, are the points

$$\alpha = \pi, \quad \alpha = 2\pi, \quad \alpha = 3\pi, \quad \ldots,$$

because the eigenvalues of the triangular kernel are π^2, $4\pi^2$, $9\pi^2$, ... as we know from § 3·10. Moreover, as Iglisch showed, these points are not only 'possible', but also actual, bifurcation points; this is by no means surprising because (since previously we put $\omega = 1$) they correspond to the well-known resonance cases of the elementary pendulum problem.

Appendix I

Algebraic Systems of Linear Equations

The theory of integral equations can be thought of as a generalization of the algebraic theory of systems of linear equations. Consequently the reader should be familiar with this theory.

This is not difficult today because such a theory can be quickly deduced from Cramer's rule and a simple basic theorem about matrices with characteristic p.† But first we recall a formula of Cauchy which gives the value of a *bordered* determinant of $(n+1)$st order

$$A' = \begin{vmatrix} a & \beta_1 & \beta_2 & \dots & \beta_n \\ \alpha_1 & a_{11} & a_{12} & \dots & a_{1n} \\ \alpha_2 & a_{21} & a_{22} & \dots & a_{2n} \\ \dots\dots\dots\dots\dots\dots\dots\dots \\ \alpha_n & a_{n1} & a_{n2} & \dots & a_{nn} \end{vmatrix} = \begin{vmatrix} a_{11} & a_{12} & \dots & a_{1n} & \alpha_1 \\ a_{21} & a_{22} & \dots & a_{2n} & \alpha_2 \\ \dots\dots\dots\dots\dots\dots\dots\dots\dots \\ a_{n1} & a_{n2} & \dots & a_{nn} & \alpha_n \\ \beta_1 & \beta_2 & \dots & \beta_n & a \end{vmatrix} \quad (1)$$

in terms of the determinant of nth order $A = \| a_{rs} \|$ and the cofactors A_{rs} of its components. To be precise, we have

$$A' = aA - \sum_{h,\,k=1}^{n} A_{hk}\alpha_h\beta_k. \quad (2)$$

In fact, if we develop the determinant A' according to its first row and the corresponding cofactors according to their first columns, we obtain

$$A' = aA + \sum_{k=1}^{n}(-1)^{k+1}\beta_k \sum_{h=1}^{n}(-1)^h \alpha_h[(-1)^{h+k}A_{hk}]$$

$$= aA - \sum_{h,\,k=1}^{n} A_{hk}\alpha_h\beta_k.$$

With the help of this formula, it is easy to obtain the theorem mentioned above about matrices with characteristic p:

† For the definition of *characteristic*, see § 2·3.

If a matrix with m rows and n columns

$$\begin{Bmatrix} a_{11} & a_{12} & \ldots & a_{1n} \\ a_{21} & a_{22} & \ldots & a_{2n} \\ \cdots\cdots\cdots\cdots\cdots\cdots \\ a_{m1} & a_{m2} & \ldots & a_{mn} \end{Bmatrix} \tag{3}$$

has characteristic $p < m$ and if a non-vanishing minor of p-th order is contained in its first p rows, then the following $m - p$ rows are 'linear combinations' of the first p, in the sense that it is possible to find constants $\mu_1^{(r)}, \mu_2^{(r)}, \ldots, \mu_p^{(r)}$ such that for every r between $p + 1$ and m, and any s between 1 and n, we have

$$a_{rs} = \mu_1^{(r)} a_{1s} + \mu_2^{(r)} a_{2s} + \ldots + \mu_p^{(r)} a_{ps} \quad (p+1 \leqslant r \leqslant m, \; 1 \leqslant s \leqslant n). \tag{4}$$

However, no such relation is possible among the elements of the first p rows, even if $p = m$.

For the proof let us suppose that

$$A^{(p)} = \begin{vmatrix} a_{11} & \ldots & a_{1p} \\ \cdots\cdots\cdots\cdots \\ \cdots\cdots\cdots\cdots \\ a_{p1} & \ldots & a_{pp} \end{vmatrix}$$

is a non-vanishing minor of pth order.† We determine the multipliers $\mu_1^{(r)}, \mu_2^{(r)}, \ldots, \mu_p^{(r)}$ by considering first the p equations of type (4) corresponding only to the values $1, 2, \ldots, p$ of s. Since the determinant of the coefficients of these equations is the non-vanishing determinant $A^{(p)}$, we obtain from Cramer's rule

$$\mu_h^{(r)} = \frac{1}{A^{(p)}} \begin{vmatrix} a_{11} \ldots a_{h-1,1} \; a_{r1} \; a_{h+1,1} \ldots a_{p1} \\ \cdots\cdots\cdots\cdots\cdots\cdots\cdots\cdots\cdots \\ a_{1p} \ldots a_{h-1,p} \; a_{rp} a_{h+1,p} \ldots a_{pp} \end{vmatrix}$$

$$= \frac{1}{A^{(p)}} \sum_{k=1}^{p} a_{rk} A_{hk}^{(p)} \quad (h = 1, 2, \ldots, p), \tag{5}$$

where $A_{hk}^{(p)}$ denotes the cofactor of a_{hk} in $A^{(p)}$.

Next we show that the previous values of $\mu_1^{(r)}, \mu_2^{(r)}, \ldots, \mu_p^{(r)}$ satisfy not only the first p equations (4), but also the following $n - p$, i.e. the equations corresponding to $s = p + 1, p + 2, \ldots, n$.

† Otherwise permute the columns until this condition is satisfied.

In fact, for such values of s we have

$$A^{(p)}(a_{rs} - \mu_1^{(r)}a_{1s} - \mu_2^{(r)}a_{2s} - \ldots - \mu_p^{(r)}a_{ps}) = A^{(p)}a_{rs} - \sum_{h,\,k=1}^{p} a_{hs}a_{rk}A_{hk}^{(p)},$$

and this is, by virtue of (2), the development of the determinant

$$\begin{vmatrix} a_{11} & a_{12} & \ldots & a_{1p} & a_{1s} \\ a_{21} & a_{22} & \ldots & a_{2p} & a_{2s} \\ \ldots\ldots\ldots\ldots\ldots\ldots\ldots\ldots\ldots \\ a_{p1} & a_{p2} & \ldots & a_{pp} & a_{ps} \\ a_{r1} & a_{rs} & \ldots & a_{rp} & a_{rs} \end{vmatrix},$$

which is zero, since all the minors of $(p+1)$st order extracted from matrix (3) vanish.

Furthermore, no relation of type (4) may occur among the first p rows, because if

$$\mu_1 a_{1s} + \mu_2 a_{2s} + \ldots + \mu_p a_{ps} = 0 \quad (s = 1,\ 2, \ldots, n),$$

then, in particular,

$$\left.\begin{array}{l} \mu_1 a_{11} + \mu_2 a_{21} + \ldots + \mu_p a_{p1} = 0, \\ \mu_1 a_{12} + \mu_2 a_{22} + \ldots + \mu_p a_{p2} = 0, \\ \ldots\ldots\ldots\ldots\ldots\ldots\ldots\ldots\ldots\ldots\ldots, \\ \mu_1 a_{1p} + \mu_2 a_{2p} + \ldots + \mu_p a_{pp} = 0, \end{array}\right\}$$

and, by Cramer's rule, this implies $\mu_1 = \mu_2 = \ldots = \mu_p = 0$, since the determinant of the coefficients is the non-vanishing determinant $A^{(p)}$.

This proves the basic theorem.

Now, let

$$\left.\begin{array}{l} a_{11}x_1 + a_{12}x_2 + \ldots + a_{1n}x_n = b_1, \\ a_{21}x_1 + a_{22}x_2 + \ldots + a_{2n}x_n = b_2, \\ \ldots\ldots\ldots\ldots\ldots\ldots\ldots\ldots\ldots\ldots\ldots, \\ a_{m1}x_1 + a_{m2}x_2 + \ldots + a_{mn}x_n = b_m, \end{array}\right\} \quad (6)$$

be any system of linear equations whose coefficient matrix has characteristic $p \leqslant m$, and suppose that a non-vanishing minor of order p is contained in the first p rows. By the previous theorem we know that for $p < m$ and $p + 1 \leqslant r \leqslant m$, there are p multipliers $\mu_1^{(r)}, \mu_2^{(r)}, \ldots, \mu_p^{(r)}$ such that

$$\mu_1^{(r)}(a_{11}x_1 + a_{12}x_2 + \ldots + a_{1n}x_n) + \ldots + \mu_p^{(r)}(a_{p1}x_1 + a_{p2}x_2 + \ldots + a_{pn}x_n)$$
$$= a_{r1}x_1 + a_{r2}x_2 + \ldots + a_{rn}x_n \quad (r = p+1, p+2, \ldots, m).$$

Consequently only two cases are possible. Either

(i) $\qquad \mu_1^{(r)}b_1 + \ldots + \mu_p^{(r)}b_p = b_r \quad (r = p+1, p+2, \ldots, m),$ $\qquad\qquad$ (7)

so that the equations following the first p are consequences of these first p and do not need to be considered; or

(ii) there exists an r_0 $(p + 1 \leqslant r_0 \leqslant m)$ such that

$$\mu_1^{(r_0)}b_1 + \ldots + \mu_p^{(r_0)}b_p \neq b_{r_0}, \qquad\qquad (8)$$

and in this case the system is *incompatible*, because the r_0th equation then contradicts the first p equations.

This shows that

(*a*) if $n = m = p$, then (6) *has always one and only one solution*;

(*b*) if $n = p$, $m > p$, *then* (6) *has one and only one solution provided that the $m - p$ equations* (7) *are satisfied*;

(*c*) if $n > p$, $m = p$, then (6) has an *infinite number of solutions*, to be precise, ∞^{n-p} solutions, because the values of $n - p$ of the unknowns can be fixed arbitrarily;

(*d*) if $n > p$, $m > p$, then (6) has ∞^{n-p} *solutions*, provided that equations (7) are satisfied.

We thus see that system (6) is always *compatible* (i.e. has at least one solution) if $m = p$, while if $m > p$ it is compatible if and only if the quantities b_r satisfy condition (7).

In particular, if $m = n$, (6) has one and only one solution for $p = n$, i.e. if the determinant of the coefficients is different from zero, while if $p < n$, it has either ∞^{n-p} solutions, or none, depending on whether the $m - p$ conditions (7) are all satisfied or not.

If the system (6) is *homogeneous*, i.e. if

$$b_1 = b_2 = \ldots = b_m = 0,$$

then there is always the *trivial solution* $x_1 = x_2 = \ldots = x_n = 0$; there are *non-trivial solutions* if and only if $p \leqslant n - 1$. These solutions can be 'parametrically' represented by formulas like (2·3-13).

Appendix II

Hadamard's Theorem

In § 2·5, we needed the following theorem of Hadamard in order to establish the convergence of the Fredholm series $D(\lambda)$ and $D(x, y; \lambda)$ under the hypothesis $|K(x,y)| < N$:

If D is a determinant of n-th order with components

$$a_{rs} \quad (r, s = 1, 2, \ldots, n),$$

then
$$D^2 \leqslant \sum_{s=1}^{n} a_{1s}^2 \sum_{s=1}^{n} a_{2s}^2 \ldots \sum_{s=1}^{n} a_{ns}^2, \tag{1}$$

and consequently we have

$$|D| \leqslant n^{\frac{1}{2}n} N^n \tag{2}$$

provided that
$$|a_{rs}| \leqslant N. \tag{3}$$

Of the many proofs of this theorem, we shall give one here which makes use of the orthogonalization process of § 3·1.†

This process is now applied, not to a system of functions, but to a system of n *linearly independent*‡ *vectors of n-dimensional space*

$$\mathbf{a}_r = a_{r1}\mathbf{i}_1 + a_{r2}\mathbf{i}_2 + \ldots + a_{rn}\mathbf{i}_n \quad (r = 1, 2, \ldots, n). \tag{4}$$

By means of a suitable linear transformation of the type

$$\left.\begin{aligned}
\mathbf{b}_1 &= \mathbf{a}_1, \\
\mathbf{b}_2 &= \mu_{21}\mathbf{b}_1 + \mathbf{a}_2, \\
\mathbf{b}_3 &= \mu_{31}\mathbf{b}_1 + \mu_{32}\mathbf{b}_2 + \mathbf{a}_3, \\
&\ldots\ldots\ldots\ldots\ldots\ldots\ldots\ldots\ldots\ldots\ldots\ldots\ldots\ldots\ldots\ldots, \\
\mathbf{b}_n &= \mu_{n1}\mathbf{b}_1 + \mu_{n2}\mathbf{b}_2 + \ldots + \mu_{n, n-1}\mathbf{b}_{n-1} + \mathbf{a}_n,
\end{aligned}\right\} \tag{5}$$

---

† F. Tricomi, 'Sul teorema di Hadamard sui determinanti', *Rev. Universidad Nac. Tucuman S.A.* **1**, 1940, pp. 297–301.

‡ If the n vectors are linearly dependent, i.e. if there are n constants $\lambda_1, \lambda_2, \ldots, \lambda_n$ (not all zero) such that $\lambda_1\mathbf{a}_1 + \lambda_2\mathbf{a}_2 + \ldots + \lambda_n\mathbf{a}_n = 0$, then the determinant D vanishes and the theorem becomes obvious. We denote the n *fundamental unit vectors* of our space by $\mathbf{i}_1, \mathbf{i}_2, \ldots, \mathbf{i}_n$.

system (4) can be carried into a similar system of *orthogonal* vectors

$$\mathbf{b}_r = b_{r1}\mathbf{i}_1 + b_{r2}\mathbf{i}_2 + \ldots + b_{rn}\mathbf{i}_n \quad (r = 1, 2, \ldots, n), \qquad (6)$$

i.e. of vectors such that

$$\mathbf{b}_r \mathbf{b}_s = b_{r1}b_{s1} + b_{r2}b_{s2} + \ldots + b_{rn}b_{sn} = 0 \quad (r \neq s). \qquad (7)$$

Just as in the case of functions, we put

$$\mu_{rs} = \frac{-\mathbf{a}_r \mathbf{b}_s}{\mathbf{b}_s^2} \quad (r = 1, 2, \ldots, n; \, s = 1, 2, \ldots, r-1), \qquad (8)$$

where the denominator cannot vanish because, if it did, a linear combination of the vectors $\mathbf{a}_1, \mathbf{a}_2, \ldots, \mathbf{a}_n$ would *vanish*, contradicting the hypothesis that the vectors (4) are linearly independent.

Moreover, we have

$$\mathbf{b}_r^2 \leqslant \mathbf{a}_r^2 \quad (r = 1, 2, \ldots, n), \qquad (9)$$

because

$$\begin{aligned}
\mathbf{b}_r^2 &= [\mathbf{a}_r + (\mu_{r1}\mathbf{b}_1 + \mu_{r2}\mathbf{b}_2 + \ldots + \mu_{r,\,r-1}\mathbf{b}_{r-1})]^2 \\
&= \mathbf{a}_r^2 + \mu_{r1}^2\mathbf{b}_1^2 + \mu_{r2}^2\mathbf{b}_2^2 + \ldots + \mu_{r,\,r-1}^2\mathbf{b}_{r-1}^2 \\
&\quad + 2\mathbf{a}_r(\mu_{r1}\mathbf{b}_1 + \mu_{r2}\mathbf{b}_2 + \ldots + \mu_{r,\,r-1}\mathbf{b}_{r-1}) \\
&= \mathbf{a}_r^2 + \mu_{r1}^2\mathbf{b}_1^2 + \mu_{r2}^2\mathbf{b}_2^2 + \ldots + \mu_{r,\,r-1}^2\mathbf{b}_{r-1}^2 \\
&\quad - 2(\mu_{r1}^2\mathbf{b}_1^2 + \mu_{r2}^2\mathbf{b}_2^2 + \ldots + \mu_{r\,\,r-1}^2\mathbf{b}_{r-1}^2) \\
&= \mathbf{a}_r^2 - (\mu_{r1}^2\mathbf{b}_1^2 + \mu_{r2}^2\mathbf{b}_2^2 + \ldots + \mu_{r,\,r-1}^2\mathbf{b}_{r-1}^2) \leqslant \mathbf{a}_r^2.
\end{aligned}$$

Now, in addition to determinant D with components a_{rs}, we consider the determinant Δ with components b_{rs}, and we observe that, as a consequence of (5), we have

$$a_{rs} = b_{rs} - \mu_{r1}b_{1s} - \mu_{r2}b_{2s} - \ldots - \mu_{r,\,r-1}b_{r-1,\,s} \quad (r, s = 1, 2, \ldots, n). \quad (10)$$

Consequently, by using repeatedly the theorem on determinants with a row or column of sums of addends, we see that the determinant D can be expressed as a sum of determinants whose first term is the determinant Δ, while the successive terms are all *zero*, since at least two of their columns are identical. Hence

$$D = \Delta.$$

But, in view of (7), and the well-known rule for the multiplication of determinants, we have

$$\Delta^2 = \begin{vmatrix} \Sigma b_{1s}^2 & 0 & \cdots & 0 \\ 0 & \Sigma b_{2s}^2 & \cdots & 0 \\ \multicolumn{4}{c}{\cdots\cdots\cdots\cdots\cdots\cdots\cdots} \\ 0 & 0 & \cdots & \Sigma b_{ns}^2 \end{vmatrix} ;$$

hence, remembering (9), we finally obtain

$$D^2 = \Delta^2 = \Sigma b_{1s}^2 \, \Sigma b_{2s}^2 \cdots \Sigma b_{ns}^2 \leqslant \Sigma a_{1s}^2 \, \Sigma a_{2s}^2 \cdots \Sigma a_{ns}^2,$$

i.e. we obtain (1).

Exercises

1. Solve directly (i.e. by the evaluation of the iterated kernels and so on) the Volterra equations of the second kind with kernels $(x-y)^m$ $(m = 1, 2, 3, \ldots)$.

2. Solve the previous equations and the one with the kernel $a + b(x-y)$ $(a, b$ const.) by means of reduction to a linear differential equation.

3. Determine the resolvent kernel of a Volterra integral equation of the closed cycle (§ 1·9) with the help of the Laplace transformation.

4. Using the previous result (Exercise 3), show that to the Volterra kernel

$$(x-y)^{-\frac{1}{2}} J_1[2\lambda(x-y)]$$

corresponds the resolvent kernel

$$-(x-y)^{-\frac{1}{2}} I_1[2\lambda(x-y)],$$

where J_1, I_1 are the Bessel and modified Bessel functions respectively of the first kind and the first order.

5. Solve the Volterra equation of the first kind with the kernel

$$1 + 4y + \tfrac{3}{2}y^2 - (4 + 3y)x + \tfrac{3}{2}x^2.$$

6. Show that in the following cases the function $D(\lambda)$ of Fredholm's equation of the second kind has the given expression

Cases	$K(x, y)$	Basic interval	$D(\lambda)$
a	± 1	0, 1	$1 \mp \lambda$
b	$g(x)$ or $g(y)$	a, b	$1 - \lambda \int_a^b g(t)\,dt$
c	xy	0, 1	$1 - \lambda/3$
d	$\sin x \sin y$	0, 2π	$1 - \pi\lambda$
e	$2e^x e^y$	0, 1	$1 - (e^2 - 1)\lambda$
f	$x + y$	0, 1	$1 - \lambda - \lambda^2/12$
g	$x^2 + y^2$	0, 1	$1 - \tfrac{2}{3}\lambda - \tfrac{4}{45}\lambda^2$
h	$xy(x+y)$	0, 1	$1 - \tfrac{1}{2}\lambda - \tfrac{1}{240}\lambda^2$

7. Verify that in the following cases the solution of the indicated Fredholm equation of the second kind (basic interval $(0, 1)$) are the indicated $\phi(x)$:

$K(x, y)$	λ	$f(x)$	$\phi(x)$
1	λ	x	$x + \lambda/[2(1-\lambda)]$
1	1	0	C (const.)
1	1	$\frac{1}{2} - x$	$\frac{1}{2} - x + C$
$x + y$	λ	x	$\dfrac{6(\lambda - 2)\, x - 4\lambda}{\lambda^2 + 12(\lambda - 1)}$
$x + y$	$-6 \pm 4\sqrt{3}$	0	$(1 \pm \sqrt{3}x)\, C$
$x + y$	$-6 \pm 4\sqrt{3}$	$1 \mp \sqrt{3}x$	$1 \pm \sqrt{3}x - (1 + \frac{3}{2}x) + (1 \pm \sqrt{3}x)\, C$

8. Solve the equations of cases (a), (c) and (e) of Exercise 6, assuming that the known functions $f(x)$ are (a) $\sec^2 x$ or $\sec x \tan x$, (c) and (e) x or e^x.

9. Using the method of § 2·3 give the solution of the Fredholm equations of the second kind with the following kernels:

$$(a)\ 1 + xy, \quad (b)\ 1 + \sin x \sin y, \quad (c)\ (1+x)(1-y), \quad (d)\ x \pm y,$$
$$(e)\ 1 + x + y, \quad (f)\ (x \pm y)^2, \quad (g)\ x^2 + xy + y^2.$$

10. Solve the Dirichlet problem for a circle (i.e. obtain the Poisson integral) using the results of § 2·7.

11. Using the method of § 3·1, orthogonalize $1, x, x^2, x^3$ in the basic interval $(-1, 1)$.

12. Assuming $0 < \xi_1 < \xi_2 < \ldots < \xi_n < 1$, orthogonalize the n step functions

$$\psi_k(x) = \begin{cases} 1 & (0 \leqslant x \leqslant \xi_k) \\ 0 & (\xi_k < x \leqslant 1) \end{cases} \quad (k = 1, 2, \ldots, n).$$

13. Evaluate the infinite series Σn^{-4} and Σn^{-6} by applying Parseval's formula with respect to the trigonometric system to the functions

$$\frac{\pi^2}{6} - \frac{\pi x}{2} + \frac{x^2}{4} \quad \text{and} \quad 2x^3 - 3x^2 + x \text{ respectively.}$$

14. Deduce the classical expansion for the functions $\pi \cotg(\alpha \pi)$ and $[\pi/\sin(\alpha \pi)]^2$ in terms of rational functions, using Parseval's formula with respect to the trigonometric system for the functions $\cos(\alpha x)$ and $\sin(\alpha x)$.

15. Using the result of Exercise 11, determine the eigenvalues and eigenfunctions of the symmetric kernel $1 + xy + x^2 y^2 + x^3 y^3$ in the basic interval $(-1, 1)$.

16. Evaluate the infinite series $\Sigma n^{-4}, \Sigma n^{-6}, \ldots$ using the triangular kernel $T(x, y)$.

17. Determine the eigenfunctions and the eigenvalues of the symmetric kernel $K(x, y) = \min(x, y)$ in the basic interval $(0, 1)$.

18. The same for the kernel $K(x,y) = \exp\{\min(x,y)\}$ using the linear differential equation
$$\phi'' - \phi' + \lambda e^x \phi = 0,$$
which can be integrated in closed form by means of Bessel functions.

19. Study the symmetric kernel
$$K(x,y) = \sum_{n=1}^{\infty} n^{-1} \sin(n\pi x) \sin(n\pi y)$$
in the basic interval $(0,1)$ and, in particular, determine the iterated kernels.

20. Using the formula
$$\int_0^{\infty} t^{\alpha-1} \cos(\xi t)\, dt = \Gamma(\alpha) \, |\xi|^{-\alpha} \cos(\tfrac{1}{2}\alpha\pi) \quad (0 < \alpha < 1),$$
show that 'Abel's equation with constant limits'
$$\int_0^1 |x-y|^{-\alpha} \phi(y)\, dy = f(x) \quad (0 < \alpha < 1)$$
has no more than *one* solution (Tricomi, *Rend. Ist. Lombardo* (2), **60**, 1927, pp. 598–609).

21. Study the Sturm-Liouville system
$$\frac{d}{dx}\left[p(x) \frac{dy}{dx} \right] + \lambda r(x)\, y = 0 \quad (y(0) = y(1) = 0),$$
especially the case $p(x) = r(x) = \exp 2kx$ $(k = \text{const.})$.

22. Study the singular Sturm-Liouville system
$$\frac{d}{dx}\left(x \frac{dy}{dx} \right) + \lambda xy = 0 \quad (y(0) \text{ finite}, y(1) = 0),$$
showing in particular that
$$\sum_{h=1}^{\infty} [j_h J_1(j_h)]^{-2} J_0(j_h x) J_0(j_h y) = -\tfrac{1}{2} \log \max(x,y),$$
where $j_1, j_2, j_3, \ldots$ are the successive positive zeros of $J_0(x)$.

23. Utilize the result of Exercise 22 and the methods of § 3·11 to evaluate approximately the first zero j_1 of $J_0(z)$.

24. Study in the manner indicated in § 3·16 the unsymmetric kernel $K(x,y) = x + \sin y$ in the basic interval $(-\pi, \pi)$.

25. With the methods of § 3·18 determine approximately the first natural frequency of a membrane of elliptic shape (Frank-v. Mises [12], **2**, p. 365).

26. Use formula (4·1-10) for $\nu = \pm \tfrac{1}{2}$ to determine the self-adjoint functions in the linear transformations
$$\mathscr{S}_x[\phi(y)] = \sqrt{\left(\frac{2}{\pi}\right)} \int_0^{\infty} \sin(xy)\, \phi(y)\, dy,$$
$$\mathscr{C}_x[\phi(y)] = \sqrt{\left(\frac{2}{\pi}\right)} \int_0^{\infty} \cos(xy)\, \phi(y)\, dy.$$

230   EXERCISES

27. Use the method of § 4·3 to solve Föppl's integral equation

$$\int_{-1}^{*1} \frac{\phi(y)}{y^2 - x^2}\, dy = f(x)$$

(Hamel [14], § 16 of the 1st ed.).

28. Show that in the case $a(x) \equiv 1$ the non-trivial solutions of Carleman's homogeneous equation are

$$C(1-x)^{\mu-1}(1+x)^{\mu} \quad \text{with} \quad \mu = \frac{1}{\pi} \operatorname*{arctg}_{(0,\,\pi)}(\lambda\pi), \quad C = \text{const.}$$

29. Utilize the results of § 4·4 about homogeneous integral equations of the Carleman type to prove that for *any* choice of the constants $c_1, c_2, \ldots, c_n$ (not all zero), the function

$$\frac{c_1 \cos \xi + c_2 \cos 2\xi + \ldots + c_n \cos n\xi}{c_1 \sin \xi + c_2 \sin 2\xi + \ldots + c_n \sin n\xi}$$

cannot be continuous in the range $0 \leqslant \xi \leqslant \pi$. (Attempt to identify the previous function with the function a of the general theory.)

30. Prove that

$$\int_{-1}^{*-k} + \int_{k}^{*1} \frac{\operatorname{sgn} y}{\sqrt{[(1-y^2)(y^2-k^2)]}} \frac{dy}{y-x}$$
$$= \int_{-1}^{*-k} + \int_{k}^{*1} \frac{|y|}{\sqrt{[(1-y^2)(y^2-k^2)]}} \frac{dy}{y-x} \equiv 0 \quad (0 < k < 1)$$

(Tricomi, *Z. angew. Math. Ph.* 2, 1951, pp. 402–6).

31. Solve the non-linear integral equation

$$\phi(x) + \lambda \int_0^1 (x+y)\, \phi^2(y)\, dy = 0$$

(Richard, *Atti Accad. Sci. Torino*, 78, 1943, pp. 293–311).

32. Study the non-linear differential system

$$y'' = \lambda \frac{x}{y}, \quad y'(0) = 0, \quad y(1) = 0$$

by transforming it into an integral equation of the Hammerstein type.

REFERENCES†

[1] d'Adhémar, R. *L'équation de Fredholm et les problèmes de Dirichlet et de Neumann.* Gauthier-Villars, Paris, 1909.

[2] Bôcher, M. *An Introduction to the Study of Integral Equations.* Cambridge Tracts, no. 10, 1909.

[3] Bückner, H. *Die praktische Behandlung von Integral-Gleichungen.* Springer, Berlin, Göttingen, Heidelberg, 1952.

[4] Carleman, T. *Sur les équations intégrales singulières à noyau réel et symétrique.* Univ. Årsskrift, no. 3, Uppsala, 1923.

[5] Collatz, L. *Eigenwertprobleme und ihre numerische Behandlung.* Akad. Verlagsgesell., Leipzig, 1946; Chelsea, New York, 1948.

[6] Courant, R., and Hilbert, D. *Methoden der mathematischen Physik,* 1. 2 Aufl., Springer, Berlin, 1931.

[7] Davis, H. T. *The Present Status of Integral Equations.* Indiana Univ. Studies, no. 70, 1926.

[8] Davis, H. T. *The Theory of the Volterra Integral Equation of Second Kind.* Indiana Univ. Studies, 17, 1930.

[9] Doetsch, G. *Theorie und Anwendung der Laplace-Transformation.* Springer, Berlin, 1937; Dover, New York, 1943.

[10] Doetsch, G. *Handbuch der Laplace-Transformation,* Birkhäuser, Basel, 1, 1950; 2, 1955; 3, 1956.

[11] Evans, G. C. *Functionals and their Applications.* Amer. Math. Soc., Coll. Publ. no. 5^1, New York, 1918.

[12] Frank, P., and v. Mises, R. *Die Differential- und Integralgleichungen der Mechanik und Physik,* 1, 2. Vieweg und S., Braunschweig, 1930–5.

[13] Goursat, É. *Cours d'analyse,* 3, 4ᵉ ed. Gauthier-Villars, Paris, 1928.

[14] Hamel, G. *Integralgleichungen, Einführung in Lehre und Gebrauch.* 2 Aufl., Springer, Berlin, etc., 1949.

[15] Hardy, G. H., Littlewood, J. E., and Pólya, G. *Inequalities.* Univ. Press, Cambridge, 1934.

[16] Hellinger, E., and Toeplitz, O. *Encykl. Math. Wiss.* 2^3, 1927, pp. 1335–1601.

† This register of references embraces only *books and monographs* on integral equations and a few related matters. The reader will find some further references (to single papers) in footnotes, but all this is far from a modern bibliography on integral equations! A good bibliography on the subject (until about 1928) is contained in the book of Vivanti [50] and embraces about five hundred papers. Another rich bibliography (of about the same size and time) is that of Davis [7]. Today a quite complete work of this kind would embrace probably more than a thousand titles and consequently would be almost useless unless very accurately organized.

[17] Heywood, H. B., and Fréchet, M. *L'équation de Fredholm et ses applications à la physique mathématique.* 3e ed., Paris, 1923.

[18] Hilbert, D. *Grundzüge einer allgemeinen Theorie der linearen Integralgleichungen.* Teubner, Leipzig, 1912.

[19] Hille, E. *Functional Analysis and Semi-groups.* Amer. Math. Soc., Coll. Publ. no. 31, New York, 1948.

[20] Hohenemser, K. *Die Methoden zur angenäherten Lösung von Eigenwertproblemen in der Elastokinetik.* Ergebnisse der Math. 1^4, Springer, Berlin, 1932.

[21] Ince, E. L. *Ordinary Differential Equations.* Longmans Green and Co., London, 1927.

22] Janet, M. *Équations intégrales et applications à certaines problèmes de la physique mathématique.* Mémor. Sciences Math. nos. 101–102, Gauthier-Villars, Paris, 1941.

[23] Kaczmarz, S., and Steinhaus, H. *Theorie der Orthogonalreihen.* Warszawa, Lwòw, 1935.

[24] Kneser, A. *Die Integralgleichungen und ihre Anwendungen in der mathematischen Physik.* 2 Aufl., Vieweg und S., Braunschweig, 1922.

[25] Knopp, K. *Theorie und Anwendung der unendlichen Reihen.* 3 Aufl., Springer, Berlin, 1931; Engl. ed., Blackie and Sons, London and Glasgow, 1928.

[26] Korn, A. *Über freie und erzwungene Schwingungen. Eine Einführung in die Theorie der linearen Integralgleichungen.* Leipzig, 1912.

[27] Kowalewski, G. *Integralgleichungen.* W. de Gruyter, Berlin, 1930.

[28] Krall, G., and Einaudi, R. *Meccanica tecnica delle vibrazioni,* 1, 2. Zanichelli, Bologna, 1940.

[29] Lalesco, T. *Introduction à la théorie des équations intégrales.* Gauthier-Villars, Paris, 1922.

[30] Lovitt, W. V. *Linear Integral Equations.* McGraw-Hill, New York, 1924; new edition s.d.

[31] Morse, Ph. M. *Vibration and Sound.* McGraw-Hill, New York, 1936.

[32] Muskhelishvili, N. I. (*Singular Integral Equations. Boundary Problems of the Theory of Functions and Certain of their Applications to Mathematical Physics.* Russian.) OGIZ, Moscow, Leningrad, 1946; English transl., Melbourne, 1949.

[33] Navarro Barrás, F. *Conferencias sobre la teoría de las ecuaciones integrales (lineales y no-lineales).* Cons. Sup. Investig. Científicas, Madrid, 1942.

[34] v. Neumann, J. *Charakterisierung des Spektrums eines Integraloperators.* Actualités scient. indust. no. 229, Hermann, Paris, 1935.

[35] Paley, R. E. A. C., and Wiener, N. *Fourier Transforms in the Complex Domain.* Amer. Math. Soc., Coll. Publ. no. 19, New York, 1934.

[36] Petrovsky, I. G. *Vorlesungen über die Theorie der Integralgleichungen.* Physica-Verlag, Würzburg, 1953.

[37] Picone, M. *Appunti di analisi superiore.* Rondinella, Napoli, 1940.

[38] Pólya, G., and Szegö, G. *Isoperimetric Inequalities in Mathematical Physics.* Ann. of Math. Ser. no. 27, Princeton, s.d., (1951).

[39]   Riesz, F. *Les systèmes d'équations linéaires à une infinité d'inconnues.* Gauthier-Villars, Paris, 1913.

[40]   Schmeidler, W. *Integralgleichungen mit Anwendungen in Physik und Technik.* Akad. Verlagsgesell., Leipzig, 1950.

[41]   Stone, M. H. *Linear Transformations in Hilbert Space and their Applications to Analysis.* Amer. Math. Soc., Coll. Publ. no. 15, New York, 1932.

[42]   Szegö, G. *Orthogonal Polynomials.* Amer. Math. Soc., Coll. Publ. no. 23, New York, 1939.

[43]   Titchmarsh, E. C. *Introduction to the Theory of Fourier Integrals.* Univ. Press, Oxford, 1937.

[44]   Titchmarsh, E. C. *Eigenfunction Expansions Associated with Second-order Differential Equations.* Clarendon Press, Oxford, 1946.

[45]   Tricomi, F. G. *Lezioni di Analisi Matematica,* 1, 2. 7th ed., 'Cedam', Padova, 1956.

[46]   Tricomi, F. G. *Equazioni differenziali.* 2nd ed., Einaudi, Torino, 1953.

[47]   Tricomi, F. G. *Serie ortogonali di funzioni.* Gheroni, Torino, 1948.

[48]   Tricomi, F. G. *Lezioni sulle equazioni integrali.* Gheroni, Torino, 1954.

[49]   Tricomi, F. G. *Vorlesungen über Orthogonalreihen.* Springer, Berlin, Göttingen, Heidelberg, 1955.

[50]   Vivanti, G. *Elementi della teoria delle equazioni integrali lineari.* Hoepli, Milano, 1916; German ed. (edited by F. Schwank), Helwingsche Verlag, Hannover, 1929.

[51]   Volterra, V., and Pérès, J. *Théorie générale des fonctionnelles,* 1. Gauthier-Villars, Paris, 1936.

[52]   Weyl, H. *Singuläre Integralgleichungen mit besonderer Berücksichtigung des Fourierschen Integraltheorem.* Diss., Göttingen, 1908.

[53]   Whittaker, E. T., and Watson, G. N. *A Course of Modern Analysis.* 4th ed., Univ. Press, Cambridge, 1927.

[54]   Wiarda, G. *Integralgleichungen unter besonderer Berücksichtigung der Anwendungen.* Teubner, Leipzig, 1930.

[55]   Widder, D. V. *The Laplace Transform.* Univ. Press, Princeton, 1946.

[56]   Wintner, A. *Spektraltheorie der unendlichen Matrizen.* Hirzel, Leipzig, 1929.

[57]   Zaanen, A. C. *Linear Analysis. Measure and Integral, Banach and Hilbert Spaces, Linear Integral Equations.* Bibl. Math., Amsterdam, 1953.

Index*

* This index does not embrace locutions like: '*integral equations*',
'*differential equations*' and so on, and classical names like Cauchy, Riemann
and so on.

A CATALOGUE OF
SELECTED DOVER BOOKS
IN ALL FIELDS OF INTEREST

A CATALOGUE OF SELECTED DOVER
BOOKS IN ALL FIELDS OF INTEREST

CONDITIONED REFLEXES, Ivan P. Pavlov. Full translation of most complete statement of Pavlov's work; cerebral damage, conditioned reflex, experiments with dogs, sleep, similar topics of great importance. 430pp. 5⅜ x 8½.   60614-7 Pa. $4.50

NOTES ON NURSING: WHAT IT IS, AND WHAT IT IS NOT, Florence Nightingale. Outspoken writings by founder of modern nursing. When first published (1860) it played an important role in much needed revolution in nursing. Still stimulating. 140pp. 5⅜ x 8½.   22340-X Pa. $3.00

HARTER'S PICTURE ARCHIVE FOR COLLAGE AND ILLUSTRATION, Jim Harter. Over 300 authentic, rare 19th-century engravings selected by noted collagist for artists, designers, decoupeurs, etc. Machines, people, animals, etc., printed one side of page. 25 scene plates for backgrounds. 6 collages by Harter, Satty, Singer, Evans. Introduction. 192pp. 8⅞ x 11¾.   23659-5 Pa. $5.00

MANUAL OF TRADITIONAL WOOD CARVING, edited by Paul N. Hasluck. Possibly the best book in English on the craft of wood carving. Practical instructions, along with 1,146 working drawings and photographic illustrations. Formerly titled *Cassell's Wood Carving.* 576pp. 6½ x 9¼.
                                                              23489-4 Pa. $7.95

THE PRINCIPLES AND PRACTICE OF HAND OR SIMPLE TURNING, John Jacob Holtzapffel. Full coverage of basic lathe techniques—history and development, special apparatus, softwood turning, hardwood turning, metal turning. Many projects—billiard ball, works formed within a sphere, egg cups, ash trays, vases, jardiniers, others—included. 1881 edition. 800 illustrations. 592pp. 6⅛ x 9¼.   23365-0 Clothbd. $15.00

THE JOY OF HANDWEAVING, Osma Tod. Only book you need for hand weaving. Fundamentals, threads, weaves, plus numerous projects for small board-loom, two-harness, tapestry, laid-in, four-harness weaving and more. Over 160 illustrations. 2nd revised edition. 352pp. 6½ x 9¼.
                                                              23458-4 Pa. $6.00

THE BOOK OF WOOD CARVING, Charles Marshall Sayers. Still finest book for beginning student in wood sculpture. Noted teacher, craftsman discusses fundamentals, technique; gives 34 designs, over 34 projects for panels, bookends, mirrors, etc. "Absolutely first-rate"—E. J. Tangerman. 33 photos. 118pp. 7¾ x 10⅝.   23654-4 Pa. $3.50

CATALOGUE OF DOVER BOOKS

THE ANATOMY OF THE HORSE, George Stubbs. Often considered the great masterpiece of animal anatomy. Full reproduction of 1766 edition, plus prospectus; original text and modernized text. 36 plates. Introduction by Eleanor Garvey. 121pp. 11 x 14¾.   23402-9 Pa. $6.00

BRIDGMAN'S LIFE DRAWING, George B. Bridgman. More than 500 illustrative drawings and text teach you to abstract the body into its major masses, use light and shade, proportion; as well as specific areas of anatomy, of which Bridgman is master. 192pp. 6½ x 9¼. (Available in U.S. only)
22710-3 Pa. $3.50

ART NOUVEAU DESIGNS IN COLOR, Alphonse Mucha, Maurice Verneuil, Georges Auriol. Full-color reproduction of *Combinaisons ornementales* (c. 1900) by Art Nouveau masters. Floral, animal, geometric, interlacings, swashes—borders, frames, spots—all incredibly beautiful. 60 plates, hundreds of designs. 9⅜ x 8-1/16.   22885-1 Pa. $4.00

FULL-COLOR FLORAL DESIGNS IN THE ART NOUVEAU STYLE, E. A. Seguy. 166 motifs, on 40 plates, from *Les fleurs et leurs applications decoratives* (1902): borders, circular designs, repeats, allovers, "spots." All in authentic Art Nouveau colors. 48pp. 9⅜ x 12¼.
23439-8 Pa. $5.00

A DIDEROT PICTORIAL ENCYCLOPEDIA OF TRADES AND IN-DUSTRY, edited by Charles C. Gillispie. 485 most interesting plates from the great French Encyclopedia of the 18th century show hundreds of working figures, artifacts, process, land and cityscapes; glassmaking, paper-making, metal extraction, construction, weaving, making furniture, clothing, wigs, dozens of other activities. Plates fully explained. 920pp. 9 x 12.
22284-5, 22285-3 Clothbd., Two-vol. set $40.00

HANDBOOK OF EARLY ADVERTISING ART, Clarence P. Hornung. Largest collection of copyright-free early and antique advertising art ever compiled. Over 6,000 illustrations, from Franklin's time to the 1890's for special effects, novelty. Valuable source, almost inexhaustible.
Pictorial Volume. Agriculture, the zodiac, animals, autos, birds, Christmas, fire engines, flowers, trees, musical instruments, ships, games and sports, much more. Arranged by subject matter and use. 237 plates. 288pp. 9 x 12.
20122-8 Clothbd. $14.50

Typographical Volume. Roman and Gothic faces ranging from 10 point to 300 point, "Barnum," German and Old English faces, script, logotypes, scrolls and flourishes, 1115 ornamental initials, 67 complete alphabets, more. 310 plates. 320pp. 9 x 12.   20123-6 Clothbd. $15.00

CALLIGRAPHY (CALLIGRAPHIA LATINA), J. G. Schwandner. High point of 18th-century ornamental calligraphy. Very ornate initials, scrolls, borders, cherubs, birds, lettered examples. 172pp. 9 x 13.
20475-8 Pa. $7.00

THE AMERICAN SENATOR, Anthony Trollope. Little known, long unavailable Trollope novel on a grand scale. Here are humorous comment on American vs. English culture, and stunning portrayal of a heroine/villainess. Superb evocation of Victorian village life. 561pp. 5⅜ x 8½.
23801-6 Pa. $6.00

WAS IT MURDER? James Hilton. The author of *Lost Horizon* and *Goodbye, Mr. Chips* wrote one detective novel (under a pen-name) which was quickly forgotten and virtually lost, even at the height of Hilton's fame. This edition brings it back—a finely crafted public school puzzle resplendent with Hilton's stylish atmosphere. A thoroughly English thriller by the creator of Shangri-la. 252pp. 5⅜ x 8. (Available in U.S. only)
23774-5 Pa. $3.00

CENTRAL PARK: A PHOTOGRAPHIC GUIDE, Victor Laredo and Henry Hope Reed. 121 superb photographs show dramatic views of Central Park: Bethesda Fountain, Cleopatra's Needle, Sheep Meadow, the Blockhouse, plus people engaged in many park activities: ice skating, bike riding, etc. Captions by former Curator of Central Park, Henry Hope Reed, provide historical view, changes, etc. Also photos of N.Y. landmarks on park's periphery. 96pp. 8½ x 11. 23750-8 Pa. $4.50

NANTUCKET IN THE NINETEENTH CENTURY, Clay Lancaster. 180 rare photographs, stereographs, maps, drawings and floor plans recreate unique American island society. Authentic scenes of shipwreck, lighthouses, streets, homes are arranged in geographic sequence to provide walking-tour guide to old Nantucket existing today. Introduction, captions. 160pp. 8⅞ x 11¾. 23747-8 Pa. $6.95

STONE AND MAN: A PHOTOGRAPHIC EXPLORATION, Andreas Feininger. 106 photographs by *Life* photographer Feininger portray man's deep passion for stone through the ages. Stonehenge-like megaliths, fortified towns, sculpted marble and crumbling tenements show textures, beauties, fascination. 128pp. 9¼ x 10¾. 23756-7 Pa. $5.95

CIRCLES, A MATHEMATICAL VIEW, D. Pedoe. Fundamental aspects of college geometry, non-Euclidean geometry, and other branches of mathematics: representing circle by point. Poincare model, isoperimetric property, etc. Stimulating recreational reading. 66 figures. 96pp. 5⅝ x 8¼.
63698-4 Pa. $2.75

THE DISCOVERY OF NEPTUNE, Morton Grosser. Dramatic scientific history of the investigations leading up to the actual discovery of the eighth planet of our solar system. Lucid, well-researched book by well-known historian of science. 172pp. 5⅜ x 8½. 23726-5 Pa. $3.50

THE DEVIL'S DICTIONARY. Ambrose Bierce. Barbed, bitter, brilliant witticisms in the form of a dictionary. Best, most ferocious satire America has produced. 145pp. 5⅜ x 8½. 20487-1 Pa. $2.25

PRINCIPLES OF ORCHESTRATION, Nikolay Rimsky-Korsakov. Great classical orchestrator provides fundamentals of tonal resonance, progression of parts, voice and orchestra, tutti effects, much else in major document. 330pp. of musical excerpts. 489pp. 6½ x 9¼. 21266-1 Pa. $7.50

TRISTAN UND ISOLDE, Richard Wagner. Full orchestral score with complete instrumentation. Do not confuse with piano reduction. Commentary by Felix Mottl, great Wagnerian conductor and scholar. Study score. 655pp. 8⅛ x 11. 22915-7 Pa. $13.95

REQUIEM IN FULL SCORE, Giuseppe Verdi. Immensely popular with choral groups and music lovers. Republication of edition published by C. F. Peters, Leipzig, n. d. German frontmaker in English translation. Glossary. Text in Latin. Study score. 204pp. 9⅜ x 12¼.
23682-X Pa. $6.00

COMPLETE CHAMBER MUSIC FOR STRINGS, Felix Mendelssohn. All of Mendelssohn's chamber music: Octet, 2 Quintets, 6 Quartets, and Four Pieces for String Quartet. (Nothing with piano is included). Complete works edition (1874-7). Study score. 283 pp. 9⅜ x 12¼.
23679-X Pa. $7.50

POPULAR SONGS OF NINETEENTH-CENTURY AMERICA, edited by Richard Jackson. 64 most important songs: "Old Oaken Bucket," "Arkansas Traveler," "Yellow Rose of Texas," etc. Authentic original sheet music, full introduction and commentaries. 290pp. 9 x 12. 23270-0 Pa. $7.95

COLLECTED PIANO WORKS, Scott Joplin. Edited by Vera Brodsky Lawrence. Practically all of Joplin's piano works—rags, two-steps, marches, waltzes, etc., 51 works in all. Extensive introduction by Rudi Blesh. Total of 345pp. 9 x 12. 23106-2 Pa. $14.95

BASIC PRINCIPLES OF CLASSICAL BALLET, Agrippina Vaganova. Great Russian theoretician, teacher explains methods for teaching classical ballet; incorporates best from French, Italian, Russian schools. 118 illustrations. 175pp. 5⅜ x 8½. 22036-2 Pa. $2.50

CHINESE CHARACTERS, L. Wieger. Rich analysis of 2300 characters according to traditional systems into primitives. Historical-semantic analysis to phonetics (Classical Mandarin) and radicals. 820pp. 6⅛ x 9¼.
21321-8 Pa. $10.00

EGYPTIAN LANGUAGE: EASY LESSONS IN EGYPTIAN HIEROGLYPHICS, E. A. Wallis Budge. Foremost Egyptologist offers Egyptian grammar, explanation of hieroglyphics, many reading texts, dictionary of symbols. 246pp. 5 x 7½. (Available in U.S. only)
21394-3 Clothbd. $7.50

AN ETYMOLOGICAL DICTIONARY OF MODERN ENGLISH, Ernest Weekley. Richest, fullest work, by foremost British lexicographer. Detailed word histories. Inexhaustible. Do not confuse this with *Concise Etymological Dictionary*, which is abridged. Total of 856pp. 6½ x 9¼.
21873-2, 21874-0 Pa., Two-vol. set $12.00

THE DEPRESSION YEARS AS PHOTOGRAPHED BY ARTHUR ROTH-STEIN, Arthur Rothstein. First collection devoted entirely to the work of outstanding 1930s photographer: famous dust storm photo, ragged children, unemployed, etc. 120 photographs. Captions. 119pp. 9¼ x 10¾.
23590-4 Pa. $5.00

CAMERA WORK: A PICTORIAL GUIDE, Alfred Stieglitz. All 559 illustrations and plates from the most important periodical in the history of art photography, *Camera Work* (1903-17). Presented four to a page, reduced in size but still clear, in strict chronological order, with complete captions. Three indexes. Glossary. Bibliography. 176pp. 8⅜ x 11¼.
23591-2 Pa. $6.95

ALVIN LANGDON COBURN, PHOTOGRAPHER, Alvin L. Coburn. Revealing autobiography by one of greatest photographers of 20th century gives insider's version of Photo-Secession, plus comments on his own work. 77 photographs by Coburn. Edited by Helmut and Alison Gernsheim. 160pp. 8⅛ x 11.
23685-4 Pa. $6.00

NEW YORK IN THE FORTIES, Andreas Feininger. 162 brilliant photographs by the well-known photographer, formerly with *Life* magazine, show commuters, shoppers, Times Square at night, Harlem nightclub, Lower East Side, etc. Introduction and full captions by John von Hartz. 181pp. 9¼ x 10¾.
23585-8 Pa. $6.95

GREAT NEWS PHOTOS AND THE STORIES BEHIND THEM, John Faber. Dramatic volume of 140 great news photos, 1855 through 1976, and revealing stories behind them, with both historical and technical information. Hindenburg disaster, shooting of Oswald, nomination of Jimmy Carter, etc. 160pp. 8¼ x 11.
23667-6 Pa. $5.00

THE ART OF THE CINEMATOGRAPHER, Leonard Maltin. Survey of American cinematography history and anecdotal interviews with 5 masters—Arthur Miller, Hal Mohr, Hal Rosson, Lucien Ballard, and Conrad Hall. Very large selection of behind-the-scenes production photos. 105 photographs. Filmographies. Index. Originally *Behind the Camera.* 144pp. 8¼ x 11.
23686-2 Pa. $5.00

DESIGNS FOR THE THREE-CORNERED HAT (LE TRICORNE), Pablo Picasso. 32 fabulously rare drawings—including 31 color illustrations of costumes and accessories—for 1919 production of famous ballet. Edited by Parmenia Migel, who has written new introduction. 48pp. 9⅜ x 12¼. (Available in U.S. only)
23709-5 Pa. $5.00

NOTES OF A FILM DIRECTOR, Sergei Eisenstein. Greatest Russian filmmaker explains montage, making of *Alexander Nevsky,* aesthetics; comments on self, associates, great rivals (Chaplin), similar material. 78 illustrations. 240pp. 5⅜ x 8½.
22392-2 Pa. $4.50

GEOMETRY, RELATIVITY AND THE FOURTH DIMENSION, Rudolf Rucker. Exposition of fourth dimension, means of visualization, concepts of relativity as Flatland characters continue adventures. Popular, easily followed yet accurate, profound. 141 illustrations. 133pp. 5⅜ x 8½.
23400-2 Pa. $2.75

THE ORIGIN OF LIFE, A. I. Oparin. Modern classic in biochemistry, the first rigorous examination of possible evolution of life from nitrocarbon compounds. Non-technical, easily followed. Total of 295pp. 5⅜ x 8½.
60213-3 Pa. $4.00

PLANETS, STARS AND GALAXIES, A. E. Fanning. Comprehensive introductory survey: the sun, solar system, stars, galaxies, universe, cosmology; quasars, radio stars, etc. 24pp. of photographs. 189pp. 5⅜ x 8½. (Available in U.S. only)
21680-2 Pa. $3.75

THE THIRTEEN BOOKS OF EUCLID'S ELEMENTS, translated with introduction and commentary by Sir Thomas L. Heath. Definitive edition. Textual and linguistic notes, mathematical analysis, 2500 years of critical commentary. Do not confuse with abridged school editions. Total of 1414pp. 5⅜ x 8½. 60088-2, 60089-0, 60090-4 Pa., Three-vol. set $18.50

Prices subject to change without notice.

Available at your book dealer or write for free catalogue to Dept. GI, Dover Publications, Inc., 31 East Second Street, Mineola, N.Y. 11501. Dover publishes more than 175 books each year on science, elementary and advanced mathematics, biology, music, art, literary history, social sciences and other areas.